The Technical Writer's Handbook

The Technical Writer's Handbook

WRITING WITH STYLE AND CLARITY

Matt Young

UNIVERSITY SCIENCE BOOKS

University Science Books
55D Gate Five Road
Sausalito, CA 94965
www.uscibooks.com

Text and jacket designer: Robert Ishi
Compositor: Asco Trade Typesetting Limited
Printer and binder: Offset Paperback Manufacturers

Library of Congress Catalog Number: 88-051695

Paperback ISBN: 1-891389-21-1

Printed in the United States of America
10 9 8 7 6 5 4 3 2 1

For my teachers,
and my students,
and all those
who unwittingly
contributed
to this book.

Contents

Preface

No passion on Earth, neither love nor hate, is equal to the passion to alter someone else's draft.

— H. G. WELLS

The prescriptivist holds that the earliest meaning of a word is the only meaning and that the language must follow certain rules as if they were revealed truth. The permissivist, by contrast, argues that language is what people say and write, and that, therefore, anything goes.

Besides permissivists and prescriptivists, there are also pragmatists. The pragmatist agrees that language must be precise, but also argues that grammar is an observational science. The pragmatist, unlike the permissivist, insists that you have to be very careful precisely whom you observe.

This book is about technical and report writing. But it is also about writing and English in general. It is written by a pragmatist with leanings in both camps: I ask for precision in writing, but I try to be flexible and have a healthy regard for usage.

My prescription for technical writing is this: It is not (or should not be) any different from other writing. Technical writing should therefore be written largely in the first person and in the active voice. Sentences should be short and simple. The technical paper, like any other, should tell a story. And the story should be written for a reader who is not already familiar with the work.

Probably in atonement for some unknown sin, I have served for five years on the Editorial Review Board of the National Institute of Standards and Technology (formerly the National Bureau of Standards), where I work. I am now chairman. When people come into my office with poorly organized or poorly prepared papers, I usually tell them three things. First, write the way you talk; then polish. No one talks in dangling participles or passive voice. Why then write that way? I suspect because science is supposed to be cut and dried and impersonal. But it is not, and I argue that if you did something, then say so.

Second, write only one thought per sentence. Finally, remember that the reader does not yet understand what is now trivial to you. Your job is therefore to explain it so that it can be understood.

I learned most of what I know about English from four of my high school English teachers, Mr. Kasper, Miss Butler, Miss Quilty, and Miss Cahill. I do not even remember all of their first names, but I am grateful for the sound training they must have given me. In addition, I learned to be precise from David MacAdam, formerly of Eastman Kodak, when he edited the first edition of my book, *Optics and Lasers*, and when I served as an editor of *An Optical Communications Glossary* with my friends Glenn Hanson, Evie Gray, and others. I sharpened my diplomatic skills when I served on the Boulder Editorial Review Board with Hassel Ledbetter, John Moulder, Bob Peterson, Jane Callanan, and others. I am especially indebted to Evie Gray, Jane Callanan, Shirley Weisz, John Moulder, and Alice Levine for their perceptive comments and thorough reading of the manuscript. Robert Day made excellent suggestions about the organization of the book, and Aidan Kelly edited the manuscript with care and precision. Finally, I am grateful to Rich Donnelly for putting me in touch with University Science Books.

In addition, I keep a number of books on my desk. These include *Words on Words*, by John B. Bremner (see Bibliography, page 231); *The Elements of Style*, by William H. Strunk, Jr., and E. B. White; and two books by Rudolf Flesch, *The Art of Readable Writing* and *The ABC of Style*. I especially like Flesch's books, though I find that — for technical writing — he errs on the side of too much informality. (Flesch was famous for a time as the author of *Why Johnny Can't Read*. *The Art of Readable Writing* has been out of print for years. I would commit anything short of larceny to get my own copy. Fortunately, Flesch's books still appear from time to time in various paperback incarnations.) Finally, for a relatively painless review of grammar and punctuation, I recommend Karen Gordon's books, *The Transitive Vampire* and *The Well-Tempered Sentence*.

When I want to check out a point of punctuation or grammar, I use *The United States Government Printing Office Style Manual*. Others will find *The Chicago Manual of Style* more suited to their needs. Since many of my own publications are in American Institute of Physics journals, I keep handy a copy of the AIP *Style Manual*, by David Hathwell and A. W. Kenneth Metzner. It is a courtesy to the editors to submit manuscripts that follow their journal's style, and I recommend that all authors be familiar with the journal to which they are submitting their papers and that they read its style manual.

How to Use This Book

This book is written in two parts. Part I is a short narrative, *Introduction to Technical Writing*, whereas Part II, *An ABC of Technical Writing*, is written in alphabetical, or dictionary, format. Many of the entries in Part II are grammatical terms like *misplaced modifier*, punctuation marks like *hyphen*, or technical terms like *SI units*. Other entries, however, are idiosyncratic in that they are words or phrases, like *end result*, *found to be* and *comprise*, that frequently appear in technical or scientific papers. Still other entries are words like *resume*, *conference proceedings*, and *introduction*. You would be unlikely to look up many of these phrases if you just used the book as a reference handbook. The book should therefore be read from cover to cover and, only then, treated as a reference handbook.

Sometimes, in the text, I refer to an author by name. A few of these are famous people, like Emerson. The others, like Flesch and Tichy, are the authors of the books cited in the Bibliography. In citing these authors, I do not refer to specific pages or chapters. These citations should therefore be regarded less as formal references than as acknowledgements of those who originated some of the ideas I use in the book.

Matt Young

The Technical Writer's Handbook

Part I

An Introduction to Technical Writing

"Begin at the beginning," the King said, very gravely, "and go on till you come to the end: then stop."

—LEWIS CARROLL

It would be very easy to show how technical or report writing differed from other writing. My purpose, however, is to stress the similarities. Writing is for communication. This is no less true of technical writing than of writing for newspapers, magazines, or books of fiction. Except perhaps for a few novelists or poets, we do not write solely to express ourselves; we write to say what we have to say, so that our audience can understand it.

Many technical writers, unfortunately, seem to forget that their intention is to communicate, and they write as if for themselves. Their papers are insufficiently explanatory, and they are written with little or no regard for style or clarity. Occasionally, I hear of an important paper that has been ignored because no one could understand it, until the work was rediscovered independently by someone else. The original paper had no function except to establish priority.

Even relatively good technical writing is frequently characterized by long, complicated sentences and difficult prose. The writing is very formal, rarely uses proper names, and almost never uses the pronouns, *I*, *you*, or *we*. Such writing has to be very good if it is to avoid being very tedious.

One of the reasons for this kind of formality is the prohibition against using *I* or *we*. I do not know where this prohibition came from; it originated in this century and often exists only in the mind of the writer. The American Institute of Physics *Style Manual*, which prescribes the style for more than a dozen technical journals, specifically advises contributors to AIP journals to use the first person. Even so, most of the authors of papers submitted to these journals strenuously avoid writing *we*.

When I sit down to write a paper, I ask myself how a writer like Hemingway would write it. Hemingway. Not Faulkner, not Joyce. Hemingway, who wrote short, crisp, punchy sentences that never left any doubt what he meant.

A Farewell to Arms begins

> In the late summer of that year we lived in a house in a village that looked across the river and the plain to the mountains. In the bed of the river there were pebbles and boulders, ...

The modern technical writer might begin it

> At that point in time, a house in a village was lived in that looked across the river and the plain to the mountains, in the bed of which one could see pebbles and boulders, ...

A parody? Of course! But the style is only a slight exaggeration of the style of many technical papers. How much better they would be if they were written in the style of the original quotation!

I know many scientists who can speak perfectly clearly and expressively. Put a pencil into their hands, and they clam up. Why? I suspect that it has to do with their belief that scientific writing has to be written in that formal and unnatural style, replete with passive verbs and misplaced modifiers, shown in the Hemingway parody. When I see a manuscript written in this style, but poorly, I always advise

---

Rule #1

Write the way you talk;
then polish.

Jacques Barzun, the linguist, points out that language is not divided into poetry and prose; there is also speech, and speech is not prose. Speech is characterized by vocal inflections, as well as by pauses, backtracking, false starts, hems, haws, and, sometimes an interminable succession of *y'knows*—not to mention misplaced modifiers and unattributed pronouns. Prose, according to Barzun, is not so characterized.

English prose is only a few hundred years old, and Barzun thinks we are still developing it. The earliest prose was complicated and flowery. Jefferson's most famous sentence, *We hold these truths to be self-evident* . . . , is 111 words long. In modern times, when most of us are in a hurry (and some may be speed readers), we cannot put up with too many 111-word sentences that begin with principles and end up overthrowing governments.

Some prose is still complicated and flowery; some is complicated and clumsy. But, in this century, many writers have discovered that good prose can be direct and simple. Their writing closely approximates speech, except that it omits the hems, haws, and *y'knows*—and it is more precise than speech. It has to be: writing lacks vocal inflections, tone of voice, and gestures. Perhaps more important, the writer can't tell when the reader does not understand.

Prose has to be written more precisely than speech, but not in a stilted or unnatural manner. You have to be careful of word order, making sure that a modifier is properly located with respect to the word or phrase it modifies, but you can still write short sentences that sound like highly polished speech. In short, good prose should read like polished speech, but not stilted speech.

That is why I always advise beginners, *Write the way you talk; then polish.* Polishing includes eliminating the obvious flaws, like slang and contractions, and you do some polishing automatically as you write. Afterward, you will want to go over the manuscript carefully, looking for sentences that are hard to understand, modifiers that are out of place, sentences that are too long, or ideas that do not follow one another and need additional clarification.

Many writers write long, complicated sentences. Sports writers are among the worst offenders, but even popular science writers are sometimes guilty of:

> Though this individual (called 0H62 because she's
> the sixty-second Olduvai hominid to be discovered)

> probably had, like other *H. Habilis* specimens, a rather modern head—with a large forehead, indicating greater brain power, and a rather delicate face—her arms dangled to her knees.

This writer has tried to give you too much information in a single sentence. Some scientists do the same thing:

> The square of the autocorrelation function of an edge-enhanced image of a manufactured object was obtained by generating a Fourier-transform hologram in a converging-beam optical correlator using a liquid-crystal display video monitor addressed by a charge-coupled device video camera, which has good rectilinearity.

This author is trying to impart information in parallel instead of in series. In a single sentence, he tries to tell us about the experiment, the object used, the type of correlator, the type of display, the type of video camera, and the reason for using it. Several short sentences would have been better, even though a bit longer:

> The purpose of the experiment was to recognize a manufactured object optically. For video input, we used a charge-coupled device, because it has good rectilinearity. We displayed an edge-enhanced image on a liquid-crystal TV monitor and placed it in the object plane of a converging-beam optical correlator. We generated a Fourier-transform hologram of the display on the monitor and used it to obtain the autocorrelation function.

The second example obeys

Rule #2

Write one thought per sentence.

You don't have to adhere strictly to this rule. The main point is not to pack too much information into too short a space. Make your writing flow, one connected thought after another, so that the reader sees where you begin, where you are going, and where you end up. Tell your story logically and completely, conveying each important fact in a single, tightly constructed sentence (or, at least, a single independent clause).

A corollary to Rule #2 is

Rule #2a

Be explicit.

Don't make the reader infer something, and don't bury an important fact where it might be overlooked. For example, if you want to tell certain details of your experimental system, don't let us infer what is important by hiding these details inside a cluster of modifiers. *The single-mode, water-cooled, HCN laser beam was directed into the blazed, evacuated, ruled-grating, infrared spectrometer.* If it is important, for example, that the laser beam was a single-mode beam or that the spectrometer was evacuated, tell us in a separate sentence, so that we don't overlook the important fact.

Other writers seem to forget that their readers do not yet know what these writers are about to tell them. They remind me of a colleague I once had; he started a seminar with the words, "Most of you have heard me ramble about these things before, so I'll just go on." (Unfortunately, I had not.) These writers start their stories in the middle and work their way toward the end or, if we are lucky, toward the beginning and the end simultaneously. They do not realize that they have to give uninformed readers enough background to educate them to the point where they can understand the new material. These writers just present their material with no explanation or background.

The better approach is

Rule #3

Write for the uninformed reader.

When I am enlisting people to give a seminar or a colloquium to a group, I always ask them to choose a level of presentation that will enable a typical college senior in physics or electrical engineering (or whatever) to understand it. I ask them to take great pains to avoid speaking only to the two or three specialists in the room and, instead, to speak to the rest of us. In other words, speak to an intelligent and sophisticated, but relatively uninformed audience. Explain. Simplify. Use analogies. Rationalize when necessary. But do not go so fast that you leave most of us behind.

Similarly, do not write your paper for the two people at your institution and the three others elsewhere who will understand it instantly. Write it for a wider audience. How wide will depend on the publication. Obviously, a very short paper written for a highly specialized conference needs less explanatory material than a longer paper written for a journal with tens of thousands of subscribers. Even so, try to write each paper so that you address as wide an audience as possible, within the constraints laid down by the publication. The result will often be an elegant paper that will be appreciated for its insight by specialists and nonspecialists alike.

A paper addressed to a wide audience will probably be read and cited more than one addressed to a narrow audience. More important, though, you will have a greater chance of being understood if you write simply and include ample background. I think, therefore, that it is preferable to err on the side of simplicity and to try to explain as much and as clearly as possible.

* * *

So far, I have addressed the similarities between technical and nontechnical writing. Everything I have said could be applied equally well to magazine writing or news broadcasting. How does technical writing differ from more general writing, and why do many technical writers have so much trouble expressing themselves?

First, the material is often extremely complex. Sometimes, alas, authors do not thoroughly understand their subject; then, writing a clear paper is almost impossible. More often, though, authors understand their area very well, but are unable to express themselves simply and clearly. Theorists, for example, may see the equations before them as in a book, but not know how to put the words in between.

The problem, in a way, is that the material has become so familiar that it is trivial. It is extremely difficult to explain trivial material to someone else. Try, for example, to tell a fifth grader why division by a fraction is the same as multiplying by the reciprocal of that fraction. You can't just say, "That's the way you do it"; that is not good enough. Rather, you have to work through the problem *as if you were seeing it for the first time*, even though that may well not be the way you first learned it.

The equation alone,

$$[1/(a/b)] \cdot [(b/a)/(b/a)] = b/a,$$

is also not good enough. You have to explain, step by step, what you did and why. For that purpose, you have to use words.

The same is true in technical writing. Start with something your audience already knows and go on from there. Give a reason for every step. For example, don't say *Linearizing the equation*, followed, with no further explanation, by the result; rather, write *Taking the logarithms of both sides, we derive a linear equation of the form . . . .*

Explain yourself every step of the way. Leave out nothing except algebra and arithmetic. Be very specific in describing a design or an apparatus. If you have to skip, say so, and outline the proof or give a very specific reference.

Technical writing is also characterized by very precise terms. This leads to complex modifiers and, frequently, clumsy constructions. For example, many digital watches and portable computers use something known as a *liquid crystal display*. When you use such a display to make a video monitor, you end up with a *liquid crystal display video monitor*. Hyphenating (*liquid-crystal display video monitor* or *liquid-crystal-display video monitor*) helps only slightly; you are still left with a clumsy string of five nouns that are hard to break up. Such constructions are common in technical writing and help make it difficult to read.

Often, we solve the problem by making up an abbreviation, such as *LCD monitor* or *LC TV*. As long as the abbreviation is familiar, that is an acceptable solution; otherwise, it may make your writing even more inaccessible. The better solution is to use prepositional phrases or subordinate clauses, *a video monitor that uses a liquid crystal display*, for example. Afterward, just refer to the monitor as *the monitor*.

Technical writing also uses made-up words and precisely defined words that may not have the same meanings as they do in ordinary speech. For example, *repeatered*, in electrical engineering, may be inelegant, but it has a specific definition: a *repeatered cable* is *a cable in which one or more repeaters have been installed*. Similarly, words like *energy*, *force*, and *power* have different meanings to the physicist than to the layman. Power has yet other meanings to the political scientist and the psychologist.

Grappling with highly qualified (or modified) nouns is a real problem in technical writing and should not be dismissed lightly. It is one of the things that makes many technical papers hard to read. It is also no accident that one of the longest discussions in this book concerns *hyphenation*.

* * *

Most technical papers are organized according to a formula. That formula is essentially the one we learned in junior high science class: Introduction, Theory, Experiment, Observations, and Conclusions (including Discussion or Recommendations). In addition, most papers have an Abstract and References, and some have one or more Appendixes. These sections may not be explicit in the paper, and one or more may be missing, but this is still a rough outline of much technical writing. It is usually a good outline to follow, whether you are writing a one-page paper, a letter to the editor, a thesis, or a lengthy report.

The American National Standards Institute suggests a similar organization: Introduction, Materials (when appropriate), Method, Results, and Discussion. This form is widely used and, in effect, combines Theory and Experiment into Methods; and Observations and Conclusions into Discussion. As above, the sections need not be explicit in the paper; both formats are simply suggested outlines to follow. Neither, for example, precludes beginning a paper with its main conclusions or recommendations, and ANSI points out that the material should not be forced to conform to the ANSI format. In particular, short letters, papers with lengthy appendixes, mathematical or theoretical papers in which the methods and the argument are the same, and engineering specifications or instruction manuals need not follow either recommended format in detail.

The Title is also an important part of a paper. In fact, it is the part that will be read by the most people. Do not skimp on your Title; make it say something to the reader. No one will read your paper if the Title is uninteresting or unintelligible. Make the Title brief and to the point, but let it tell the reader that an interesting paper or a meaningful result follows. A poor Title or a Title that does not convey the real importance of your paper usually sells your work short.

The Abstract is equally important. Too many authors write a significant paper and then toss off an Abstract that conveys little or no information. Today, when there are Abstract journals and computer databases, many readers will see nothing but your Abstract. Give them enough information to make them want to go to the trouble of finding your paper and reading it. Conversely, give them enough information that they can avoid looking up your paper only to find that it is irrelevant to them. In short, make your Abstract an informative summary of your paper and include your results or recommendations whenever possible.

The Introduction is the place to give the background for your paper. You may want to explain why you pursued the line of research you did, or you may want to describe the work that led to your own. Don't make the Introduction a series of references that tell who did earlier work but nothing about that earlier work. To the greatest extent possible, let your paper stand on its own feet, so that it can be read without recourse to the References. Only those who want to study the subject in greater detail should have to consult the References.

The Introduction can also include a summary of the paper. There is nothing wrong with telling what you are going to demonstrate in the body of the paper and, perhaps, how you are going to do it. Complex papers with relatively narrow interest can often be improved greatly if the Introduction is so complete that the nonspecialist can read it alone and profit from it.

ANSI recommends, further, that you make your assumptions explicit in this section and justify your method when there is an alternative. Even the specialist has not followed your line of reasoning before; that is why you are writing the paper.

Some papers contain both a Theory section and a section on Experiment or Apparatus, whereas others have only one of these sections. This is where you get down to business and, often, write solely for the specialist. Describe your experimental design or your theoretical approach in as much detail as you think necessary, but try not to leave out whole lines of thought (without saying so). Make your paper detailed enough that an expert can reproduce your experiment or your calculation without reinventing most of your development. If the description or the theory seems too tedious, consider putting some of the more detailed or mathematical material into an Appendix.

As a general rule, write a section on Discussion or Results separate from your detailed Theory or Experiment section. A reader might initially want to skip the more complicated parts of a paper and yet learn of your new results. Whenever possible, explain your results so that the reader can appreciate them without reading the entire paper in detail.

Include an explicit Conclusion section only if there is still something unsaid, or if you think that the conclusions should be grouped together and stressed. But do not feel that you must have a Conclusion, and then write a vague summary with no real point to it. Make every word count.

References are for a reader who wants greater detail or additional background. They are not a substitute for an Introduction, nor are they a list of credits, as at the end of a movie. Make your References as specific as possible. If the publisher allows it, include titles and first and last pages of papers you cite, and include page or chapter in book citations.

* * *

Complex material and highly precise definitions are, I think, the major characteristics that distinguish technical writing from other writing. Other differences include the use of figures and tables, references, and equations. But these differences are not as fundamental, because they do not influence technical writing style directly.

This book is largely a plea for simplicity, for what Flesch calls *a word diet for those who are verbally overweight.* I recommend writing short sentences in the active voice, rather than long sentences in the passive voice. I argue in favor of a short word or phrase in place of a long or pompous one. I suggest using the pronouns *I* and *we* whenever possible.

In grammar, I demand precision, but I am not a purist or a prescriptivist. I avoid the dangling participle and insist on the correct use of adverbs. But I see nothing wrong with many split infinitives, and I begin sentences with *It is* when I think it is appropriate. I try to make my writing as close to colloquial as I can, without crossing the line into inelegance.

This book is for those who love writing or who want to improve their writing skills, but it is aimed most specifically at scientists and engineers, and most of the examples are drawn from scientific papers. Lawyers, administrators, and others who frequently write formal letters, reports, or briefs could also use it profitably by skimming over the entries that are specific to scientific writing. The book covers the gamut from letter writing to publications in archival journals. It is written in an alphabetical format, and it includes entries on grammar, usage, and definitions, as well as other entries specific to scientific and technical writing. (Because of the format, some concepts may be repeated and a few errors pointed out more than once.) The grammar entries themselves are not exhaustive; rather, I have chosen only those that seem most germane to technical and report writing. I assume, in other words, that the reader knows the rudiments of grammar and punctuation, and I concentrate instead on those areas with which technical writers have the most difficulty.

I have included many examples both of writing that could stand to be improved and of my suggested improvements. Most of the examples have been gleaned from real papers and manuscripts, but I have shortened many or edited them so that they display only one error at a time. To those who think they recognize their own handiwork, I can only apologize and offer the hope that the author was someone else.

The examples appear in *italics*. Cross references are in **boldface**. Some of the entries contain commas, so cross references are separated by semicolons, not commas.

Despite its format, this book is intended to be read from cover to cover, casually, during lunch hours or short breaks, the way I have read several other similarly organized books. My purpose is to make the case for clarity and informality in technical writing.

Many of the entries are words or phrases that appear frequently in technical writing. I have chosen these entries with care, to highlight the problems in a way that listing them under a general heading would not. Therefore, the owner who does not read the book from cover to cover will ultimately overlook a lot. After a first reading, the book should prove useful as a reference handbook.

As poet and etymologist John Ciardi used to say, Good words to you!

Part II

An ABC of Technical Writing

Most men revere words they cannot understand and consider a writer whom they can understand to be superficial.

—ALBERT EINSTEIN

A

a, an. The indefinite articles. It is archaic to use *an* before an *h* word like *historian*, since nowadays we pronounce the initial *h*. See also **abbreviations, acronyms; the**.

a while, awhile. This must be two words when used as the object of a preposition:

> The patient talked for a while but then withdrew.

When it is used as an adverb, however, it may be either one word,

> The patient talked awhile but then withdrew,

or two,

> The patient talked a while but then withdrew.

In the last example, the preposition *for* is understood. See also **all right**.

abbreviations, acronyms. Technical writing is full of abbreviations, perhaps too full. Avoid making up abbreviations just to save space; use words instead. For example, the statement

> It is clear that DFW has more effect on SNR than DFR

is not at all clear if you have just invented the abbreviations, DFR and DFW. Instead, if SNR is a standard term for signal-to-noise ratio, rewrite the sentence as

> Defocus error when writing has more effect on SNR than defocus error when reading.

When an abbreviation is common and everyone will understand it, however, use it. There is nothing wrong or inelegant in using abbreviations. But some abbreviations, like *op-amp* for *operational amplifier*, *lab* for *laboratory*, or *specs* for *specifications*, are **slang** and should be avoided in writing, even if they have been defined earlier. Similarly, avoid using an adjective, such as *Dewar*, in place of a noun phrase, *Dewar flask*, unless that usage is almost universal.

Many abbreviations are initials, like *RNF* for *refracted near field.* These abbreviations are commonly capitalized, even when they do not stand for a proper noun. Periods are usually omitted from common abbreviations. Some journals abbreviate *ac, dc, ir,* and perhaps a few other common phrases with lowercase letters; check their style manuals.

When an abbreviation, like *NATO*, is pronounced as a word, it is known as an *acronym.* Acronyms are usually capitalized when the letters are the initials of a proper noun. Other acronyms, like *radar*, are not capitalized, particularly when their origin has been mostly forgotten. When in doubt, check a **dictionary** or a **style manual**.

Which indefinite article—*a* or *an*—do you use before an abbreviation? That depends on the first letter of the abbreviation. I usually choose the article according to how I pronounce the sentence: *a NATO officer*, but *an NBS scientist*, since I read the second example, *an en bee ess scientist.*

Many publications abbreviate and capitalize *Eq. (1)* or *Fig. 3.* When you begin a sentence with *Equation* or *Figure*, however, spell the word fully. See also **SI units**.

about. *About* or **roughly** are perfectly good synonyms for *approximately.*

above, below. These are adverbs and very clumsy when used as nouns or adjectives, as in the phrases, *In the above we argued that* . . . or *In the above discussion* . . . . *The discussion above* . . . would be better, but best might be simply, *We argued above that* . . . or *We argued in the preceding section that* . . . .

absolute construction. A phrase that usually precedes a sentence and has no obvious grammatical connection to the sentence. For example,

> The grating aligned, the spectroscope is now ready for use.

Here, *the grating aligned* is an absolute construction or absolute phrase and does not, like an ordinary modifier, have to modify *spectroscope.* Beware, however, of the **dangling modifier**, which looks dangerously like an absolute construction. (The difference is that a dangling modifier can grammatically be taken to modify the subject, even when logically it should not.)

Absolute constructions may be considered *sentence modifiers* (see **adverb**) and must be set off in commas. They do not require **being**,

> The grating being aligned, the spectroscope is now ready for use,

and, indeed, are stylistically better without it.

Often, an absolute construction is weak and does not make your point forcefully enough. For example,

> Gravitational calculations on an ice sheet are much simpler, the ice sheet being relatively flat and homogeneous

stops short of telling *why* the calculations are much simpler on an ice sheet. The sentence would have been better with *because*:

> Gravitational calculations on an ice sheet are much simpler, because the ice sheet is relatively flat and homogeneous.

absolute words. Certain words, such as *unique*, *infinite*, *certain*, *perfect*, *essential*, and *necessary*, are *absolute words*. These cannot be intensified with modifiers. For example, something is either perfect or it is not; it cannot be *very perfect*.

Unique, in particular, is often misused. It is not a synonym for *very unusual*. Something is unique only if there is nothing else like it; it cannot be *very unique*. Unique should not be modified, except perhaps for *almost unique*, which might apply if there are only two or three similar things, and *truly unique*, which shows that you are using the word correctly.

Infinite is another absolute word. As *unique* should not be intensified, *infinite* should not be diminished or qualified with modifiers. What, after all, does *nearly infinite* mean? One million? One billion? If you mean that something is so large that it can be considered infinite for practical purposes, write just that, but do not write

> Long-term memory is essentially infinite in capacity.

Similarly, do not write

> When anthrax is inhaled or ingested, it can be intensely lethal.

Anthrax is either lethal or it is not. The author probably meant

> When anthrax is inhaled or ingested, it can kill quickly.

absolutely. Avoid using *absolutely* to intensify **absolute words**. Instead of

> The fact of evolution is absolutely certain; only the mechanism is in doubt,

write

> The fact of evolution is certain; only the mechanism is in doubt,

or, shorter,

> Evolution is a fact; only the mechanism is in doubt.

Either something is certain or it is not; you cannot intensify *certain* by making it *absolutely certain*. Similarly, avoid *absolute fact*; a statement is a **fact** or it is not. See also **redundant expressions**.

abstract. The abstract is one of the most important parts of any paper and will probably be read by far more people than will ever read your paper; this is especially so nowadays, when a computer search may yield no more than a **title** and an abstract. Therefore, put as much about your paper into your abstract as you possibly can. Too many abstracts are written as if they were tables of contents, with the phrases *is described, is discussed* added to each entry. Such abstracts are known as *descriptive abstracts*; they are generally not informative enough. They may tell what subjects are in the paper, but they tell nothing about the details, much less the conclusions, of your paper. They tease, rather than inform.

Consider, for example, this excerpt from a hypothetical abstract:

> Saturable dyes for laser Q switches are found to degrade with time. The cause of this degradation is discussed and a possible solution is proposed.

This example gives me no useful information, and requires me to find and read the paper.

In contrast, compare with an *informative abstract* such as

> Saturable dyes for laser Q switches degrade with time. The blue light from the flashlamp causes the degradation of the dye, and the dye's life may be extended by replacing the cell's windows with red filters.

This example is a paraphrase from an abstract I read at least ten years ago and would have forgotten had it not been so clear and informative.

Make your abstracts informative, not descriptive. That is, make your abstracts short summaries of your paper, including especially any conclusions you may have drawn or recommendations you may have made. A good informative abstract may state the problem you are examining or the procedure you have carried out, your purpose in doing it, your results, and your conclusions or recommendations. Most abstracts are limited to one paragraph and must therefore be constructed with care.

Where possible, avoid narrowly technical terms, uncommon **abbreviations**, and **references**. Make sure that the abstract stands alone and can be read without reference to the paper itself.

access. In computer jargon, *access* is properly used as a verb. For example, you can *access* a file on a disk or a subroutine from a main program. But avoid using the verb *to access* in any other context, such as

> Interested parties may access the laboratory during working hours.

Gain access to is gobbledygook for *enter*.

accuracy. See **precision and accuracy**.

acid test. If you have been holding this book in your right hand and using your left thumb to browse through it from the end to the beginning, you may think I am going to say that this is a redundant synonym for *test*. It is not. The *acid test* (from the test for gold) is not just *any* test but *the crucial* test. See, however, **redundant expressions**.

acknowledgements. This is the proper place to thank all the people who helped you, read your manuscript, and gave you their unpublished results; a **private communication** in your list of references is almost valueless. You should seek permission before you publish someone else's unpublished results and acknowledge them in writing.

I have, incidentally, spelled *acknowledgement* with a *ge*, since I learned the rule that a *g* is hard unless it precedes an *e* or an *i*. Conditioned reflex dies hard. The more common U.S. spelling is without the *e*. If you are submitting an article to a journal that will typeset your article, spell it their way.

acronym. See **abbreviation, acronym**.

action plan. Sometimes called *plan of action*. What other kind of plan is there? A plan of inaction, I suppose, is when you are planning to take a nap. In all other contexts, *action plan* is redundant and should be simply *plan*. See also **redundant expressions**.

active voice. Too many technical papers are written in the **passive voice**, like this sentence. Perhaps some technical writers like the passive voice because it allows them to avoid the **first person**. *In this paper a generalization is presented that* ... would be better rewritten as *This paper presents a generalization that* ... or *In this paper, I present* ... For one thing, the word order is better when the subordinate clause beginning with *that* directly follows the modified word *generalization*. See **misplaced modifier**.

Sentences written in the active voice are usually easier to read and slightly shorter than those written in the passive voice. In addition, the active voice often gives your paper a less clumsy or less impersonal tone. (Although many fine technical writers write elegantly in the passive voice, most cannot.) See also **dangling modifier**.

adaptation. Not *adaption*.

adjective. A word (or, sometimes, a group of words) that modifies a noun. Generally, an adjective should not be used alone, in place of a noun, though sometimes such use becomes idiomatic, as in *flat* for *flat glass*, *Dewar* for *Dewar flask*, or *schematic* for *schematic diagram*. On the other hand,

> The temperature approaches critical

is much better as

> The temperature approaches the critical temperature.

The *noun-as-adjective* (or *attributive noun*) construction is quite common in scientific writing, and I once saw it derided as *the adjective noun use tendency*. Although this phrase was intended to question the use of nouns as adjectives, it is also a parody intended to point out the tendency to string too many attributive nouns or adjectives in a row.

The adjective noun use tendency is especially troublesome in **titles**. Some authors try to pack so much information into their titles that the titles almost become abstracts. A title contained the longest unbroken string of nouns I have ever seen:

> Air Force ring laser gyro mirror investigation.

With hyphens, it would have been only slightly better.

Long strings of adjectives and nouns should be broken up with prepositions and, yes, possessives. A noun collapses under the weight of too many modifiers. Usually one or more can be cut or removed to the arm's length of a prepositional phrase. The cumbersome title,

> Continuous-wave mode-locked Nd:YAG-pumped
> subpicosecond dye laser,

could well have been shortened to

> Continuously mode-locked, laser-pumped dye laser.

If you count *dye laser* as a unit, this form has three modifiers, *continuously, mode-locked*, and *laser-pumped.* Not counting prepositional phrases, that is about all the modifiers a noun can stand, even if one of them is an adverb. Three other egregious examples found, not in titles, but embedded in sentences:

> chrome-oxide-on-chrome-on-glass layers,
>
> post-weld-heat-treatment specimen wide plate test results,

and

> an uncoated 125 micrometer diameter fine wire Pt/Pt-10% Rh thermocouple.

Make what you can of those! I call them *polymodifiers.*

In general, if you have to use two hyphens in a single modifier, you are, at best, dangerously close to a polymodifier. For example,

> The intensity was determined by the excitation-process-determined atomic displacements

is very clumsy and hangs four modifiers onto the noun *displacements.* The sentence would be much clearer rewritten

> The intensity was determined by the atomic displacements caused by excitation processes.

An exception to the two-hyphen rule is when the modifier involves a prepositional phrase, like *signal-to-noise ratio.* You could probably get away with *Middle-to-Upper Cretaceous shales,* but don't go overboard:

> They found reduced carbon in Middle-to-Upper Cretaceous organic-rich shales.

This sentence, too, should be rewritten with a prepositional phrase, for example,

> They found reduced carbon in organic-rich shales from the Middle-to-Upper Cretaceous period.

See also **hyphen**.

advance planning. I never knew anyone who did his planning in arrears, although I can think of people who do not plan at all. Since all planning is done in advance, this is a **redundant expression**.

advantage. A word often used in **circumlocutions**.

> Such a design has the advantage that it minimizes heat flow

or

One advantage of such a design is that it minimizes heat flow

should usually be shortened to

Such a design minimizes heat flow.

advent. Means *coming, arrival,* or *beginning,* as in *the advent of spring.* It does not mean *invention.* Yet *the advent of the laser* has become such a hackneyed phrase that it has acquired a life of its own. You will never catch me using it, though. See also **cliches**.

adverb. What ever happened to **-ly**? Don't write *near-diffraction limited lens* or *triple-clad fiber.* Use the adverbs *nearly* and *triply,* without a **hyphen**.

Dropping the *-ly* and using the adjective is becoming common and is not always wrong. *First, second,* and *third,* not to mention **thus**, are already perfectly good adverbs and do not require the ending *-ly.* Still, using an adjective to modify another adjective almost forces you to use hyphens and makes your writing look like **jargon**.

Adverbs and adverbial phrases can also be used as *sentence modifiers.* These frequently precede the sentence,

Luckily, no one was present when the gas escaped,

or are set off with commas,

No one was present, luckily, when the gas escaped.

Some writers regard phrases that begin with *assuming that, provided that, based on, according to,* or *depending on* as sentence modifiers. Too often, however, these phrases have all too much in common with **dangling modifiers**, and I recommend that they be avoided. For example,

Assuming that the components are in place, the laser is brought into alignment

contains, to my ear, a dangling participle, and I would rewrite the sentence

When the components are in place, the laser is brought into alignment,

or

> Assuming that the components are in place, we bring the laser into alignment.

In the last example, the phrase *assuming that* ... modifies the subject *we*. It is, however, harder to argue with the sentence

> Provided that the components are in place, the laser may be aligned,

because *provided that* could not easily be used in an adjectival phrase; therefore the phrase need not be considered dangling. I would therefore accept the initial phrase as a sentence modifier, though I am not sure all authorities will agree.

adverse, averse. These are not synonyms. *Averse* means *opposed to, reluctant*, whereas *adverse* means *unfavorable*. People are *averse* to something; the weather is *adverse*. Mark Twain was wrong, by the way, when he said that no one does anything about the weather. I do something about adverse weather: I dress for it.

advice, advise. *Advice* is the noun; *advise* is the verb. *Advise* means *give advice* and is not a synonym for *inform*. Do not write

> This letter is to advise you that your warranty has expired

when you mean

> This letter is to inform you that your warranty has expired

adviser. The preferred spelling.

advised of. Told.

affect, effect. These words are often confused, and I do not know any easy way to distinguish them. Both can be used as verbs, but (except for a technical term in psychology) only *effect* can be used as a noun: *cause and effect, the effect was brought about by* ....

Affect is more often used as a verb; it means *to have an effect on*, as in

The gravitational field of Jupiter affects the motion of the spaceship.

Effect can also be used as a verb, and it means *cause* or *bring about*, as in

We can effect a change in the motion of the spaceship by firing the rockets.

Sometimes either word may seem to fit, and the meaning may become unclear:

Firing the rockets effects the desired orbit change

could be misunderstood to mean *affects*, or *influences*, the orbit change, not *brings it about*. It would be better to rephrase the sentence, for example, by using *brings about* for the more formal *effects*:

Firing the rockets brings about the desired orbit change.

affix. The people at the Post Office dug up this silly word so that they could write the bureaucratic phrase, *affix postage*, in place of the perfectly clear phrase, *put stamp here*. Avoid *affix*; use *fasten* or *attach*.

aforementioned. What a pompous word! Leave it to the members of the legal profession. See also **above, below; false elegance; vague words, vogue words**.

agenda. Each item in a list of things to be discussed is an *agendum*. The plural is *agenda*; hence the whole list is now called an *agenda*, and the word is considered singular. Then why isn't *data* singular? Don't ask me. It just isn't. See **Ciardi's law**.

agreement. See **singular or plural**.

ain't. I see nothing wrong with *ain't*, but it ain't acceptable in good company. I find it a lot less barbaric than *aren't I?* which is generally considered acceptable.

albeit. An archaic synonym for *although*. Avoid it.

all right. Not *alright*, in spite of *almost, already*, and *always*.

allude. An *allusion* is an oblique reference; to *allude* is to refer indirectly. *Allude* is not a synonym for *refer* and should not be used as such.

alone together. Can't be!

although, though. As far as I can tell, these are synonyms, though *although* is more common at the beginning of a sentence.

am in receipt of. Got or have received. See also **business letter**.

among, between. At one time, you used to do something *between* two people and *among* three or more. That distinction seems to be fading, and *between* is replacing *among* almost universally, so you will hear on the news (even on National Public Radio News) *an agreement between the United States, Britain, and Canada.* In my writing, I try to preserve the distinction, but I do not harp on it.

By the way, since *between* is a preposition, *between you and me* is correct, not *between you and I.*

amount. Often put in redundantly, especially in the phrase *amount of time*.

> The problem is the amount of time it takes to make the measurement

can easily be shortened to

> The problem is the time to make the measurement.

Similarly, don't write *amount of loss* or *amount of mass*, for example, when *loss* or *mass* will serve.

analog. The spelling *analogue* is obsolete; if you prefer this spelling, be consistent with *catalogue, dialogue*, and so on.

and (comma), but (comma). *And* and *but* are conjunctions; in principle, they should be used, for example, to join two independent clauses. But you can begin a sentence with either of these conjunctions if you want to make a sentence stand out and yet emphasize the connection with the previous sentence. They are not, however, synonyms for *in addition* or *however* and should not be followed by a **comma**.

> And, we determined the coefficient of expansion to be $3 \cdot 10^{-6}$

should be written without the comma or with *And* replaced by *In addition*. See also **plus**.

and so. Usually **so** is enough.

> The method does not converge, and so the sequence was truncated

would be just as good without *and.* It would also be all right written with a semicolon:

> The method does not converge; therefore the sequence was truncated.

and/or. Why use such a clumsy locution? Instead of *a car and/or a bicycle,* use *a car or a bicycle, or both.* See also **slash.**

ansatz. A German word that scientists or mathematicians sometimes use for *trial solution*. The word is not in any English dictionary I checked. If you are writing in English, use English words. See also **foreign words and phrases**.

antenna. The Latin **plural** *antennae* is incorrect in engineering, and I would discourage its use in biology as well. Use the English plural, *antennas*.

anthropogenic. Rather formal synonym for **man-made**.

anxious, eager. These are not synonyms. *Anxious* implies waiting with anxiety, whereas *eager* connotes looking forward to something. I *eagerly* await your arrival, not *anxiously*, unless, perhaps, you are a hangman.

any and all. Usually a synonym for *any*, so use *any*. See also **each and every**.

any more, any time. Two words, unlike *anywhere*. One clue whether to omit the space and join the two words is the pronunciation.

Anywhere, anyone, and *anybody* are accented on the first syllable, whereas *any more* and *any time* are accented on the second. We generally do not join word pairs until their pronunciation changes so that the first syllable is accented.

Incidentally, *any more* is not a synonym for *nowadays*.

We don't do it that way any more

means we used to do it that way, but now we don't. The confusing construction

Any more, we do it the other way

is coming to mean

Nowadays, we do it the other way.

Nowadays (which, by the way, is accented on the first syllable and is therefore one word) is preferable when you mean *now, at present*, or *in these times*. Use *any more* only in negative sentences.

aperture. Not aperature.

apostrophe. Some writers subscribe to a strange rule that if a singular noun ends in *s*, you form the possessive by adding only an apostrophe, not *'s*. The word is supposed to be pronounced as if the *'s* were there, but, because the *s* is omitted in writing, you often hear people incorrectly failing to pronounce the final *s* as well. Some books extend this rule to words that end in *z* and even *x*.

I consider the rule silly and put *'s* on all singular nouns (if the editor will let me). True, this sometimes leaves you with *Moses's commandments*, but so what? You can always write *the commandments of Moses* instead. In any case, even if you write *Moses' commandments*, you are supposed to pronounce all three esses.

When a name ends in *s* (*Pockels, Huyghens*), do not butcher it by inserting an apostrophe before the final *s*: *a Pockels cell*, not *a Pockel's cell; Huyghens's principle* or *Huyghens' principle*, not *Huyghen's principle*. Incidentally, there seems to be no rule for using the possessive for principles, quantities, or devices named after people. For example, common usages are *a Pockels cell, a Michelson interferometer, a Skinner box; Poisson's ratio, Young's modulus, Gauss's law; Keynesian economics, Gaussian curve, Cartesian coordinates*. There is no consistency, so use usage as your guide. Do not write *the Young modulus* when everyone else writes *Young's modulus*; it just sounds strange.

The possessive of a plural noun is formed by adding the apostrophe alone, and the missing *s* is not pronounced. When a plural does not end in *s*, the possessive is formed by adding *'s*, as in *children's* and *women's; womens'* is incorrect.

When the possessive applies to several nouns connected by *and* or *or*, generally attach the *'s* to the last noun only:

> Michelson and Morley's experiment was the most famous experiment ever to yield a null result.

Sometimes, especially with pronouns, you might want to make both pronouns or a noun and a pronoun possessive. For example,

> I shall illustrate with examples from Marie's and my research

just sounds better than

> I shall illustrate with examples from Marie and my research,

even though the research was done jointly and the author did not mean his research and Marie's research.

The plurals of dates and acronyms may be written with or without an apostrophe: *1940's, UFO's*, or *1940s, UFOs*. I prefer these plurals with the apostrophe, and some **style manuals** require it. If you use periods in an **abbreviation**, you almost have to use an apostrophe:

> Five of the participants were M.D.s

was very hard to read and would have been better as

> Five of the participants were M.D.'s.

Possessive pronouns never take an apostrophe: *its, theirs*, never *it's, their's*. (*It's* is a **contraction** of *it is*, never the possessive of *it*.)

appendix. Some wag once called an appendix a vestigial portion of a book, for which no one has yet discovered a function. Many appendixes—those that try to give you all there is to know about light rays in three pages, for example—are useless. The author should have devoted a full chapter to light rays or told what you have to know before reading the book.

Still, there is often a function for an appendix in a technical book or paper: it is often the best place to put detailed calculations, computer programs, or long lists of numbers. Especially in a complex theoretical paper or in a paper with a lot of experimental detail, you may want to tell your story briefly in the main body of the paper, so that it is accessible to a somewhat general audience. Put the fine detail into an appendix for the reader who is interested in the specifics.

application of. Applying. See **-ion, -ity**.

appositive. A noun is in *apposition* to another when you write something like

> The aerospace giant, McDonnell-Douglas, bid on the project.

The appositive, *McDonnell-Douglas*, is set off in commas. You would, however, omit the commas in the sentence

> Aerospace giant McDonnell-Douglas bid on the project.

Here *Aerospace giant* is practically a modifier, so the noun, *McDonnell-Douglas*, is not set off in commas.

In technical writing, most journals avoid commas when the appositive is a **symbol**. Thus,

> The angle u subtended by the disk is 0.707 times the numerical aperture of the system

is preferred over

> The angle, u, subtended by the disk is 0.707 times the numerical aperture of the system.

The appositive u by the way, is in apposition to the noun *angle*, not *disk*, so the phrase should not be written

> The angle subtended by the disk, u, is 0.707 times the numerical aperture of the system.

Numbered equations, those given their own lines, are often best treated as appositives and punctuated with commas:

> We may calculate the intensity,
>
> $$I = E^*E, \tag{3}$$
>
> where * denotes complex conjugation.

When a sentence begins with an appositive, the word in apposition must be the subject of the sentence, not some modifier of the subject. For example, in

> The scientific director, Condon's remarks were inexcusable,

or the closely related construction,

> As the scientific director, Condon's remarks were inexcusable,

the leading phrases are intended to modify *Condon*, whereas grammatically they modify the subject, which is *remarks*. Neither sentence is correct; they should be rewritten

> Since Condon was scientific director, his remarks were inexcusable

or

> The scientific director, Condon, made inexcusable remarks,

depending on meaning.

An appositive can also dangle from the end of a sentence:

> One million people are infected with the virus that causes AIDS, a number that continues to grow.

AIDS, however, is not a number; it is a disease. The sentence should be rewritten as either

> One million people are infected with the virus that causes AIDS, a disease that continues to spread

or

One million people are infected with the virus that causes AIDS; their number continues to grow.

See also **dangling modifier; equations; misplaced modifier; semicolons; SI units**.

approximate. This adjective is occasionally misplaced or used where the adverb would be better.

The approximate nearfield intensity was 1 W/cm^2

should really be written

The nearfield intensity was approximately 1 W/cm^2,

since it is the value that is approximate, not the intensity.

I consider *about* and *roughly* perfectly acceptable in place of *approximately*, especially when the approximation is not so good.

artifact. Originally any object made by humans, *artifact* now means, in addition, any structure, such as a glitch in a graph, that is an artificial result of the procedure, of an error, or of an accident. The alternative spelling *artefact* is gaining acceptance in some social sciences, but I can see no justification for it. Even the etymology (from the Latin *ars, artis*) argues in favor of the *i* rather than the *e*.

as. A poor synonym for *because*, because it can also mean *while* or *at the same time that*.

As we were measuring the current, the voltage dropped.

Does this sentence mean

While we were measuring the current, the voltage dropped,

or does it mean

Because we were measuring the current, the voltage dropped?

Either meaning could be inferred. Do not rely on context to resolve the ambiguity; avoid *as* in this context and do not be afraid to use *because*. See also **because, since**.

As to is often superfluous. In

> There is no doubt as to what he meant,

delete *as to*:

> There is no doubt what he meant.

as, like. *As* is a conjunction, and *like* is a preposition, so you are not supposed to say (as I said this morning)

> He doesn't exactly sound like he has been influenced by Stravinsky.

Flesch argues, though, that this use of *like* is so common that it has now become acceptable. I agree, but I would not write it, only say it.

as can be seen. Many authors refer to a figure and then write

> As can be seen from the figure, the visible spectrum divides into two parts.

A construction that used the **active voice** in the dependent clause would be much smoother and perhaps clearer:

> As the figure shows, the visible spectrum divides into two parts.

A more direct approach, however, might be best:

> The figure shows that the visible spectrum divides into two parts.

as many as. Do not use *as many as* or **up to** to intensify a number:

> As many as 2000 people have died of AIDS since the disease was first diagnosed in the United States.

Just report your statistic without telling us to be impressed by it:

> Two thousand people have died of AIDS since it was first diagnosed in the United States.

Instead, reserve *as many as* for cases where there is doubt about the number you are reporting:

> Perhaps as many as 2000 people have died of AIDS since the disease was first diagnosed in the United States.

as much as or more than. Not *as much or more than*; do not forget the second *as* in this phrase. For example,

> The social sciences need support as much if not more than the physical sciences

or

> The social sciences need as much support if not more than the physical sciences

could be rewritten in any of several ways, depending on meaning:

> The social sciences need support as much as or more than the physical sciences,

or

> The social sciences need as much support as the physical sciences, if not more,

for example.

as noted above. If it is right above, omit this phrase; otherwise, tell exactly where it is noted. See also **above, below**.

as such. This is not a synonym for *therefore*. It should be used only when the pronoun *such* refers to a previous noun. For example,

> This laser is a powerful source of energy; as such it has proved useful for welding

contains a correct use of *as such*, because it means

> As a powerful source of energy, the laser has proved useful for welding.

It is not, however, correct to say

> Holography has been around for more than 20 years; as such it is surprising that it has not found more widespread success.

In this sentence, *as such* is used incorrectly as a synonym for *therefore*. The pronoun *such* does not refer to anything in the first clause and therefore should be replaced by *therefore*:

Holography has been around for more than 20 years; therefore it is surprising that it has not found more widespread success.

As such is sometimes close in meaning to *in itself*:

We are discussing psychoanalysis as such, not psychotherapy in general.

Here, *as such* is important because it intensifies that we are discussing psychoanalysis and nothing else.

as to whether. Whether. See also **if, whether**.

as yet. Can usually be shortened to *yet*. For example,

The technique may rejuvenate brain tissue by an as-yet unknown mechanism

is a little tighter as

The technique may rejuvenate brain tissue by a yet-unknown mechanism.

Similarly, when *as yet* is used as an adverb,

We have not as yet performed the analysis,

yet seems a bit better:

We have not yet performed the analysis.

assume. Sometimes we write as if our assumptions influenced nature.

If we assume axial symmetry, n is independent of k.

Not true! n is (or is not) independent of k no matter what we assume. The writer really meant

If we assume axial symmetry, we may assume that n is independent of k.

Is this the origin of the anthropic principle (the principle that says the universe exists *because* there is intelligent life in it)? See also **obeys an equation**.

assuming. Not a synonym for *if*.

This is not surprising, assuming symmetry is weak

would have been better as

> If symmetry is weak, the result is expected

or perhaps

> We assume that symmetry is weak and are not surprised by the result.

Some writers use *assuming* or *assuming that* as a sentence modifier; see **adverb**.

asterisk. Not pronounced asteriCK.

at this juncture. Now. Not, incidentally, *at this junction.*

at this point in time. Whether or not you believe that time is one of four dimensions, use *now*. *At this time* and *at this point* are also lengthy synonyms for *now*.

atomic mass, molecular mass. In certain fields, mass has historically been expressed in terms of the proton mass. In **SI units**, the correct unit is the *unified atomic-mass unit*, which is designated u. 1 u is equal to (1/12) of the mass of the nucleus of carbon-12. Masses of large molecules, for example, should be expressed as 1000 u, not as 1000 daltons. A molecule whose mass is 1000 u should probably be described as having a *relative molecular mass* of 1000 u, since its mass is given relative to 1 u, approximately the proton mass. The terms *atomic weight* and *molecular weight* are incorrect and should be avoided. See also **mass, weight**.

attributive noun. See **adjective; inanimate possessive**.

author. Never write *the author* as if you were someone else. See **first person**.

In addition, I consider *author* a lousy verb. You don't *author* a paper; you *write* it.

B

back up. The verb is two words; the adjective is either *back-up* or *backup*. Never hyphenate the verb. See **hyphen**.

bacteria. The **plural** of *bacterium* and an exception to the rule to use English plurals. Don't write

> In a bacteria that would not happen

but rather

> In bacteria that would not happen

or

> In a bacterium that would not happen.

barbarism. This is Bremner's word for an outrageously bad word or phrase. I put *prioritize, impact on, debrief, intuit, ruggedize*, and a lot of other "verbifications" into that category. In addition, I once saw *formularize* in print.

Orwell gives five rules—in essence, never use a cliche, never use a very long word, never use an unusual word, never use the passive, and cut whenever possible. He concludes with the advice to break any of these rules rather than write something "outright barbarous."

based on. Some authors use *based on* as a *sentence modifier*; because *based on* can also introduce an adjectival phrase, I prefer to avoid this use. Therefore, I would not write

> Based on data for population, disease was the major factor,

but rather

> Based on data for population, the study showed that disease was the major factor.

In the revised sentence, *based on* correctly modifies *study*, since the study is indeed based on data for population. It might be clearest, though, to avoid *based on* entirely and write

> Data for population showed that disease was the major factor.

See also **adverb; dangling modifier**.

Baumeister's law. You can't insult people by telling them something they already know. In other words, when writing a paper or, especially, when giving a talk, begin at the beginning. You can certainly leave people behind by leaving out something they do not know. This is especially important when you are preparing for a talk; here the audience is more or less captive and cannot stop to check your references when it does not understand something. If you err at all, err on the side of review, simplicity, and clarity. It does not matter how much material or how much depth you present if no one understands you.

became aware of. Heard; learned.

because, since. *Since* is a synonym for *because*, but sometimes there may be confusion because of its other meaning, *from the time that*. For example,

> Since the electricity failed, we have had no heat.

Does this mean *because* the electricity failed, or does it mean *from the time that* the electricity failed? The sentence is ambiguous, and the reader should not have to rely on context to understand it.

Because should also be used in place of cumbersome phrases like *due to the fact that* and *as a result of the fact that*. The phrase *This is because* is clumsy; it is better written *This is so because*.

being. Often put into a sentence superfluously, as in

> The devices generated two voltages, one being proportional to the *E* field and one to the *H* field.

Here, *being* could just as well have been left out:

> The devices generated two voltages, one proportional to the *E* field and one to the *H* field.

Being that is sometimes used as a synonym for *because*. You find it more in speech than in writing:

> Being that we had the day off, I did not go into the lab

is arguably correct, but weak. In writing, use

> Because we had the day off, I did not go into the lab.

A related construction is

> That form of leukemia being caused by a virus, they reasoned that all leukemia may be caused by a virus.

This is not quite a dangling participle, but it is nevertheless barely acceptable. Here also, better to use *because*:

> Because that form of leukemia is caused by a virus, they reasoned that all leukemia may be caused by a virus.

See also **absolute construction; dangling modifier**.

believe, feel, think. These are not synonyms, though you often hear them used interchangeably. I *believe* in human rights, but I *think* they are not always respected, and I *feel* bad about that. I *believe* in determinism, so I *think* that quantum mechanics is not the final theory. If you can logically substitute *think* for *believe* or *feel*, then *think* is the word you want.

In a sentence like

> We believe that the theory will extrapolate well,

use *think*, *hope*, or *anticipate* in place of *believe*, depending on your meaning. Likewise, in

> The material is believed to have high thermal capacity,

change *believed* to *thought* or, better, write

> We think that the material has high thermal capacity.

below. See **above, below**.

better than. Avoid *precision better than.* Use *greater than* or *higher than.* Higher precision is not better unless you need it.

between. Write *between 5 and 10 percent* or *5–10 percent*, not *between 5 to 10 percent* or the hybrid *between 5–10 percent.* In a typescript, either make sure that no one could confuse your hyphens with minus signs or entirely avoid using hyphens with numbers. See also **among, between; from; hyphen; numbers in sentence; SI units.**

big of, good of. As in *a pretty big of a pulse.* This usage must be by analogy with *much of* and *some of*, but *much* and *some* are nouns and *big* is an adjective. Thus, the correct usage should be *a pretty big pulse* or, more formally, *a relatively big pulse.* Likewise, it is not good style to write *very intensive of capital*, although *capital intensive* is certainly idiomatically correct.

black hole. A noun that has collapsed under the weight of too many modifiers. See **adjective**.

brackets. Sometimes called *square brackets* [ ] to distinguish them from *curly brackets* { }. Outside mathematics, brackets are used mostly for **references** and for inserting a word or phrase into a quotation written by someone else,

> "Fusion, like fission, produces a substantial amount of [radioactive] waste," he said.

Here, the word *radioactive* was not in the original quotation.

You might also use brackets if you do not like to nest parentheses:

> (In the year 1905 [his *annus mirabilis*], Einstein practically founded three fields of physics.)

In addition, I always find brackets jarring when they surround a single letter. For example, you need not use brackets when you change a small letter in the original to a capital, as at the beginning of a sentence. *[S]everal days ago* ... can be written *Several days ago* ... without doing any injustice to the source.

breezy. I see no reason that a technical paper cannot be written in a somewhat breezy style, if by breezy we mean with short sentences, short words, and simple expressions. That would be a great improvement over the murky prose we sometimes see. I think it is important, though, not to get too colloquial (even in a **conference proceedings** paper) and to maintain a serious tone—serious, as columnist Russell Baker would say, but not solemn. See also **sentence length**.

Buckley's law. Never use a short word when polysyllabic terminology will suffice.

bureaucratese. See **gobbledygook**.

business letter. Avoid canned (and pompous) phrases like *I am in receipt of, enclosed please find, thank you for your kind consideration*, and so on. Each of these phrases has a translation into plain English: *I have received, here is, many thanks for your help*, and so on. Just write the business letter in plain words, the way you would talk to someone, and then perhaps polish it a bit.

Likewise, when you are finished with whatever you have to say, just stop writing. If you need to thank the recipient, avoid

> Thank you very much for your kind consideration to this matter

or some similarly pompous phrase. Instead, thank the recipient specifically:

> Many thanks for your help!

or

> Thanks very much for taking care of the problem for me.

Incidentally, Flesch says never write

> Thank you very much in advance.

If something is worth thanking someone for, thank that person afterward, in writing. Or, to paraphrase Sam Goldwyn, a verbal thank you is not worth the paper it is printed on. See also **cliche; stock phrases**.

but. Not a synonym for **however**. *But* is a conjunction often used to join two clauses:

> Glass has a relatively low melting point, but its viscosity near the melting point is high.

If you want to emphasize the comment about the viscosity, you might want to capitalize *but*:

> Glass has a relatively low melting point. But its viscosity near the melting point is high.

This construction is technically incorrect, since *but* is a conjunction, but you can often use it to good effect. I would not, however, accept

> Glass has a relatively low melting point. But, its viscosity near the melting point is high.

Instead, I would use *however*:

> Glass has a relatively low melting point. Its viscosity near the melting point, however, is high.

See also **and (comma), but (comma); comma**.

by. Expressing dimensions is sometimes difficult and results in a clumsy construction. *A 1 cm × 2 cm* rectangle is awkward, and so are *a 1 cm-by-2 cm rectangle*, *a 1-cm-by-2-cm rectangle*, and *a 1 × 2 cm*^{2} *rectangle*. *A 1 × 2 cm rectangle*, however, seems perfectly clear, since the dimension of the first number is implicit, as it is in *a distance of 10–20 km*, and this is the form I prefer. Still, you can often throw out these constructions and write *a square 1 cm on a side* or *a rectangle with dimensions 1 cm by 2 cm*. See also **hyphen**.

C

calibrate out. As in

> We calibrated out these error terms.

An unusual, but not unclear, way of saying

> We compensated for these error terms with our calibration.

The sentence

> We calibrated these error terms

does not have the same meaning.

can be seen to be. A long way to say *is*, as in

> From Fig. 3 the slope of the line can be seen to be 1/2.

A better way to write the sentence is in the **active voice**,

> Figure 3 shows that the slope of the line is 1/2.

cancel out. Cancel.

cannibalize. A cannibal is a creature that eats its own species. Nevertheless, the term *cannibalize* is accepted to mean *salvaging parts from old or obsolete equipment*.

cannot. One word, not *can not*.

capitalization. This is sometimes a problem in **titles**. Some journals capitalize only the first letter of the first word of a title:

> Imaging without mirrors or lenses: the pinhole camera and its relatives

Others capitalize only the first letter of each line of titles that exceed one line:

> Imaging without mirrors or lenses:
The pinhole camera and its relatives

Look at an issue of the journal or check the appropriate **style manual**.

In general, however, titles are written with both capital and lowercase letters:

> Imaging without Mirrors or Lenses: The Pinhole Camera and Its Relatives

All words except articles and prepositions begin with capital letters. Some authors make the error of failing to capitalize participles like *using*; these are not prepositions and should be capitalized. I would not, however, capitalize *due to* and perhaps not *owing to*, since these are compound prepositions.

When you compile any sort of bibliography or a list of **references**, choose your own capitalization style and use it consistently, irrespective of the style of the original papers. Writing some titles with caps and some without is inconsistent.

In text, use capitals for proper nouns or **trade names** and, of course, for *I* and the first words of sentences. Capitalize *Fig. 3, Table 2, Eq. (17), Sample 5*, and the sections of your paper, *Introduction* and *Section 4*, for example. Don't capitalize words like *president* and *professor* unless you are using them as proper nouns. That is, write *the professor* and capitalize the word only when it is used as a title, *Professor Hercher*, or as a form of address, *May I ask you a question, Professor?*

For specialized capitalization rules, see **abbreviations, acronyms; chemical elements; SI units**.

caption. See **figure captions**.

case. Often superfluous, as in

> For the case of measuring low energy beams, we used a longer integration time.

This clause should be rewritten

> When measuring low energy beams, we used a longer integration time.

Similarly, *for the case where* can often be replaced by *when*. For example, rewrite

For the case where x is a constant, the integration is fast

as

When x is a constant, the integration is fast.

See also **in many cases; vague words, vogue words**.

case-by-case basis. We have no policy and do not intend to formulate one.

catalog. The spelling *catalogue* is obsolete; see also **analog**.

cease and desist. Stop.

centered dot. Use the *centered* or *raised dot* for multiplication,

$$A = 40 \cdot 20,$$

and for the product of **SI units**,

$$1\ \mathrm{J} = 1\ \mathrm{W} \cdot \mathrm{s}.$$

For example, write 10 N · m, not 10 N-m and especially not 10Nm or 10N-m (without the space). I find the raised dot most commonly forgotten just before a square root sign, as in $\mathrm{MPa} \cdot \sqrt{\mathrm{Hz}}$.

International usage, as defined by the General Conference on Weights and Measures, also permits a space, 10 N m, or a period, 10 N.m, for the product of SI units. In the U.S., however, the American National Standards Institute insists on the centered dot.

cf. The abbreviation for the Latin word, *confer*, compare. Maybe OK in notes and bibliography, but I would never use it in text. See **foreign words and phrases; references**.

CGS units. Although the centimeter, the gram, and the second are perfectly good units, the system in which they are the base units is obsolete. In a few specialized fields, such as theoretical physics, it may sometimes be appropriate to use CGS units, but, in general, they should be avoided. That is, CGS units with special names should be replaced by their SI equivalents. If your field still commonly uses some CGS unit and you want to state the value of a quantity in that unit, please use the abbreviations and other recommendations of the **SI units**. In addition, put the CGS unit in a secondary position, in parentheses, for example, after the SI unit. Doing so will enable people not in your field to read your paper more easily. See also **English units**.

TABLE CGS-1. CGS units and conversions.

Name	Symbol	SI equivalent
erg	erg	10^{-7} J
dyne	dyn	10^{-5} N
poise	P	0.1 Pa · s
stokes	St	10^{-4} m^2/s
gauss	G	10^{-4} T
oersted	Oe	$1000/(4\pi)$ A/m
maxwell	Mx	10^{-8} Wb
stilb	sb	10^4 cd/m^2
phot	ph	10^4 lx

chemical elements. The names, argon, potassium, carbon dioxide, of the chemical elements and compounds are not capitalized, except, for example, when the word begins a sentence. The symbols, Ar, K, CO_2, for those elements and compounds, however, are always capitalized.

Ciardi's law. Language behaves the way it does because it does.

circumlocution. Using more words than are necessary to express a thought clearly. Many of the entries in this book, such as *in view of the fact that* and, often, *compared to*, are circumlocutions. Some scientific writers beat around the bush before they get to the point:

> In comparison with silicon-deoxidized welds, this weld did not have an improved strength-toughness relationship.

Particularly when comparing one thing with another, begin your thought with the subject of the sentence and use **than**:

> This weld did not have a better strength-toughness relationship than silicon-deoxidized welds.

Many circumlocutions begin with a dangling or nearly dangling prepositional phrase:

> For those vendors and employees who remained after the purge, the company decided it would treat them better

should be shortened to

> The company decided to treat better those vendors and employees who remained after the purge.

Similarly,

> It could be that the borders of the cleavage facets are grain boundaries

would be more direct as

> The borders of the cleavage facets could be grain boundaries.

See also **it**.

Other circumlocutions use vogue words like *property* or *result*:

> One property of a severely defocused optical system is that its MTF has isolated zeroes

should be rewritten

> The MTF of a severely defocused optical system has isolated zeroes.

A superb circumlocution that I found in a photography magazine needs no comment:

> There is this one white cat that my daughter has that I have photographed three times.

See also **vague words, vogue words; wordiness**.

cliche. Avoid cliches like the plague; one cliche in your writing could be the tip of an iceberg. And now, without further ado, see also **stock phrases**.

close proximity. Proximity.

collective noun. A *collective noun* is one like *Congress.* The Congress is the sum of all its members. Perhaps this is one reason it cannot make a decision about anything. Or do I mean *they* cannot make a decision? In U.S. English, collective nouns take the singular, whereas in British English they take the plural: Congress *is* in session, but Parliament *are* in session. In U.S. English, do not write

> The faculty has made their decision.

Instead, write either

> The faculty has made its decision

or

> The members of the faculty have made their decisions,

depending on your meaning.

I try to be precise and use the singular verb with a collective noun. What is important, though, is clarity. Once I read a discussion of the sentence

> The group were half-naked.

The writer was arguing that it was not important whether to use *was* or *were* but rather to be clear about one point: Were half the members of the group naked, or were all the members of the group half-naked? See also **singular or plural**.

colloquial. This is not a pejorative term, but merely refers to words used more in speech than in formal writing. Writing is not exactly crystallized speech, but I think nevertheless that it should be a good deal more colloquial than it is. This applies to technical writing as to any other, and a good technical paper may well be written in an informal style without crossing the line into slang or inelegance.

The contraction *it's* and the pronoun **you** are colloquial, and I use them in this book. Flesch says that they make your writing more personal and readable. I would, however, generally avoid them both in a formal paper. See also **apostrophe; its, it's**.

colon. Avoid a colon where it interrupts the flow of a sentence. For example, there is no need for the colon in

> The parts of a sentence are: subject and predicate.

The colon would make you pause unnecessarily if you were reading aloud. Similarly, there is no need for a colon before a vertical **list** or before an **equation**. These uses of the colon make your writing look forbidding. Equations especially should be treated as parts of sentences and punctuated accordingly.

Sometimes, when one sentence follows directly from another, they are separated by a colon: this use of the colon is considered elegant. The word following the colon is not capitalized in this case, since the colon has a function very similar to a semicolon.

You might, however, capitalize the next word when the next sentence begins a completely different thought.

> We performed the experiment in the following way:
> First, we prepared the sample.

Here, I would risk being called inconsistent and capitalize *First*.

When the colon is followed by words or phrases (a list, for example) that do not form a complete sentence, always use the lower case:

> We used a large number of materials: oils, waxes,
> and a few inorganic materials.

comma. Some writers, put commas, in a lot of places, where they don't belong. I call these writers, *comma-kazes*, reading their writing, and their **run-on sentences**, leaves me, comma-tose.

Commas are used sparingly these days. Use them, for example, to set off flavoring words or phrases, or **transitional devices**, like *for example* in this sentence. Use them to separate two independent clauses,

> The laser oscillated, and then it damaged a mirror,

but not a compound predicate that has only two verbs,

> The laser oscillated and then damaged a mirror.

The difference between these two sentences is *it*.

A comma is not so out of place in a compound predicate joined by *but*:

> The laser oscillated, but damaged a mirror

may not be strictly correct, but the conjunction *but* signals a change of gears that makes the comma acceptable and possibly desirable.

A comma should not be used after a *that* clause:

> Impurities that shorten the lifetime, reduce the output.

The comma is superfluous; either the clause should be set between commas

> Impurities, which shorten the lifetime, reduce the output

or the comma should be omitted. There is, however, a difference between the two meanings. In the first sentence, only those impurities that shorten the lifetime reduce the output; in the second sentence, impurities (which, incidentally, shorten the lifetime) reduce the output. A couple of commas can make a large difference in meaning. See **that, which**.

Commas are also used to separate the verbs in a compound predicate with three or more verbs, or items in a **list**. Should you place a comma before the *and* that precedes the last item? I usually do, but I don't think it really matters, as long as you are consistent throughout your paper. Do not, however, use a comma with a list that has only two items separated by *and*.

Here is an example of a comma put in for clarity at the writer's discretion:

> They wanted an air pillow between the driver's chest and head, and the steering wheel.

It would not have been correct to write

> They wanted an air pillow between the driver's chest, head, and the steering wheel.

The use of *and* and the comma makes the sentence clear, though it might be best to write

> They wanted an air pillow between the steering wheel
> and the driver's chest and head.

See also **principle of parallelism**.

Don't forget the commas that set off an explanatory or nonrestrictive phrase, especially one that defines a term or an abbreviation. For example,

> Books are subjected to vapors of diethyl zinc or DEZ
> in a pressure cooker

is ambiguous and could imply that books are subjected to vapors of either diethyl zinc or DEZ. The intended meaning is conveyed properly in

> Books are subjected to vapors of diethyl zinc, or
> DEZ, in a pressure cooker,

because the nonrestrictive phrase *or DEZ* is set off by commas.

In the US, the digits in longer numbers have traditionally been grouped in threes and the groups separated by commas:

> 1,000,000.

In technical publications that use the metric system, however, spaces are preferable to commas:

> 1 000 000.

Omit the space in a four-digit number: 1000, not 1 000, unless the number appears in a **table** with longer numbers and the space is necessary for appearance. See also **SI units.**

commence. Begin.

community. Often a wasted word. *Of interest to the optical communications community* adds nothing to *of interest in optical communications.*

comparative, superlative. This is a formal grammarian's way of saying you use *-er* for two things and *-est* for three or more. Thus, it is wrong to say *the biggest of the two* or *the greatest half*; both these constructions should use the *-er* or *comparative* form, rather than the *-est* or *superlative* form.

compare, contrast. You *compare* two things when there is a similarity; you *contrast* them when there is a difference.

compared to. I may be imagining it, but I think I see *compared to* and *compared with* more and more frequently. They are sometimes used in place of *as opposed to* but more often simply as a clumsy substitute for **than**.

> Compared to the first system, the second has fewer components

is not nearly as good or as clear as

> The second system has fewer components than the first.

Likewise, do not write

> Compared to conventional microscopy, confocal microscopy had higher resolution

or

> Spatial resolution was higher for confocal microscopy compared to conventional.

Write instead

> Confocal microscopy had higher spatial resolution than conventional microscopy.

Compared to is not a good substitute for *rather than, in contrast*, or *as opposed to*:

> Forty MCu were released, compared with 15 Cu at Three Mile Island

does not make the contrast you want. Better to use *as opposed to*:

> Forty MCu were released, as opposed to 15 Cu at Three Mile Island

or, better, write

> Forty MCu were released; in contrast, Three Mile Island released only 15 Cu.

Other sentences should be completely rewritten:

> An angle of 85° exhibited an increase in damage threshold compared to normal incidence

is harder to understand than

> The damage threshold at 85° was higher than at normal incidence.

In another variation, *advantages in comparison to other methods* should have been rewritten *advantages over other methods*. In

> The setup includes special modifications compared with the ordinary apparatus,

compared with should have been replaced by *of*:

> The setup includes special modifications of the ordinary apparatus.

See also **relative to**.

complementary. Not complImentary when you mean *completing* or *making up what is lacking.*

> Support for synchrotron light sources, as well as the complimentary neutron radiation sources, is in decline

should have used the spelling, *complementary*, since the author meant that both sources *complement* each other. A *compliment* is an expression of praise.

comprise. This word is probably misused about half the time. (The misuse will become accepted usage only when the majority of educated users accept it.) It is not a synonym for *compose* but for *is composed of.* Therefore, you cannot say *is comprised of.* The correct use of *comprise* is in a sentence like *The whole comprises the sum of its parts.*

> A board of medical examiners, comprised of some of the most respected scientists, was given authority

should have been

> A board of medical examiners, composed of some of the most respected scientists, was given authority

or, better,

> A board of medical examiners, which included some of the most respected scientists, was given authority.

Comprise is not quite the same as **include**. The object of *comprise* should be the whole, whereas the object of *include* is only a part. For example,

> The board of medical examiners comprised 15 respected scientists, including three Nobel prize winners.

Here, we may assume that there were precisely 15 members of the board, all of them respected scientists.

computer experiment. An interesting expression, which describes, for example, a series of simulations designed to discover the best or most likely value of some parameter. Not exactly theory and not exactly experiment, the computer experiment may become the third branch of science.

conceivable. When people write *It is conceivable that . . . ,* I wonder what they are trying to hide. Use *possibly.*

conclusions. Make sure your conclusions are, in fact, conclusions; often they are just a summary of the paper and do not highlight the results. If you feel you have already treated the subject adequately and have nothing more to say, just stop. There is no reason you have to have a Conclusions section.

When you elect to write a Conclusions section, state your conclusions firmly. (If they are tentative conclusions, identify them as such or use words like *probably* or *we think.*) For example, don't begin your conclusions

> We have used photothermal deflection to measure the loss in a waveguide.

If we have read the paper, we already know that. Instead begin

> Photothermal deflection is a promising technique for measuring the loss in a waveguide.

Then go on to support your conclusion.

In a **business letter** as well, writers often seem to feel they must have a concluding sentence and end up with a barbarism like *Thank you very much for your kind attention to this matter.* Why? If you have nothing more to say, just stop.

conference proceedings. I sometimes think all conference presentations should be required to begin with a joke, no matter how bad. This is your chance to express yourself! You are standing up there before 100 people! Why drone on in a monotone about klystrons, polytrons, megatrons, and waitrons when you could make them all laugh?!

Please forgive me for the **exclamation points**; we all have a little Tom Wolfe in us, no matter what Strunk and White say.

Conference proceedings are usually one of two types—longish **abstracts** or full papers. When a long abstract is called for, make sure that it gives information to a reader who has not attended the conference—it may be the only record of your presentation. Therefore, it is especially important to make the abstract precise and informative, including numbers when appropriate.

The full paper can be a lot more informal and less solemn than a paper aimed at an archival journal. I find it convenient to write the paper as if it were a transcript of my spoken presentation and therefore make my proceedings papers quite informal.

Take care, though, if you decide to read the paper rather than present it without a script. There are certain advantages to reading a paper; for example, clarity and precision are more or less guaranteed, and you will not be likely to forget anything if it is right in front of you. You may, however, find it difficult to read from a prepared script with anything like normal inflection in your voice, and this could make your presentation a disaster.

I am in the minority of presenters. Since I claim that writing is just polished speech (or should be), I usually write my paper first, more or less in transcript form. I then get a doubly or triply spaced draft and *unpolish* it—that is, make it more colloquial. I publish the original but read the unpolished version. I am, however, careful to study the paper until I am familiar enough with it that I can read it almost as if my talk were extemporaneous.

confirm or deny. A silly way of saying *comment on.*

consensus. General opinion. *Consensus of opinion* and *general consensus* are wrong.

contractions. Don't use 'em in technical writing, but use 'em in letters and informal writing.

convolve. In mathematics, the verb is *convolve*, not *convolute*. When two functions have been convolved, they are not *convoluted*, even though the result is a *convolution*. *Convoluted* is normally reserved for *convoluted arguments*, *convoluted surfaces*, or *convoluted sentences*.

cooler temperatures. The temperatures do not get cooler; the weather does. The temperatures get *lower*.

copy. *I'll copy you* does not mean *Let's play follow the leader*, but is a barbarism for *I'll send you a copy*. *I'll copy you one*, with the transitive verb, is only slightly better.

copyright. Most published material is subject to the copyright laws. You may secure a copyright on published material simply by writing the word, *copyright*, a *c* with a circle around it, the year, and your name: Copyright © 1987 by M. Young, for example. If a paper has not been published, you still may have statutory copyright, so most journals require you to transfer your copyright to them before they will publish your paper.

Publications of the United States government are not subject to copyright (though some publications with both government and nongovernment authors may be copyrighted). You are therefore free to reproduce most government publications in whole or in part without regard to copyright. (In legal jargon, they are in the *public domain*.) For example, you may reproduce a figure from an uncopyrighted paper without securing permission from the author, though it would be a courtesy to ask anyway.

The *fair use* provision of the copyright law allows teachers or librarians, for example, to make copies of articles for educational purposes. A teacher might be allowed to reproduce and distribute a small portion of a book under the fair use provision. Likewise, a reviewer or a scholar may quote small portions of a book.

With these exceptions, you must obtain permission from the copyright holder before republishing copyrighted material; this includes both text and figures. Photocopying more than a single copy of a whole article usually entails sending a fee to the Copyright Clearance Center, as detailed in the Information for Contributors section in most journals.

correlation. Correlations are usually *between . . . and*, rather than *of . . . and.*

> There is good correlation of ultrasonic velocity and formability

is clearer if rewritten as

> There is good correlation between ultrasonic velocity and formability.

could care less. Could *not* care less.

could have, might have. Not *could of, might of*, but I saw these forms in a manuscript. *Might could* and *used to could* are interesting regional expressions, but I don't think I would use either one in writing.

criterion. This is the singular; *criteria* is the plural. A *criterium* is a bicycle race.

A criterion is a rule or a standard. It is not quite correct to write

> The field strength is given by some criterion.

The field strength is, rather, given by some *value*. The *statement* that the field strength is given by that value is the criterion. The sentence should be rewritten

> The field strength is chosen according to some criterion.

critique. As a verb, this sounds like awful **jargon** to me, but it is certainly acceptable in some circles. It is not exactly the same as *criticize*, since that often has the connotation of finding fault. A critique is more a careful analysis than it is a criticism.

crosshatched. A *hatch* is a short line. This is not *crosshatched*: ///// it is *hatched*. This is *crosshatched*: XXXXX.

cubic centimeter. No longer abbreviated cc., but rather cm^3. Nowadays, the cubic centimeter and the milliliter are equivalent, though in the past they were defined slightly differently. See also **CGS units; SI units.**

cumulative. Not ACcumulative.

cut off, cutoff. The noun and the adjective are one word: *cutoff*. The verb is two words: *cut off*. See also **hyphen**.

D

dangling modifier. English nouns have no case endings, so modifiers have to be located near their nouns. In particular, when a modifier begins a sentence, it must modify the subject of the sentence. Often, however, a modifier modifies nothing in the sentence; such a modifier is known as a *dangling modifier*. In technical writing, most dangling modifiers are *present participles*, or verb forms that are used as adjectives and end in *-ing*.

> Using a collimating lens, the *xy* plane is thrown to infinity.

The *xy* plane does not use a collimating lens; you do. The sentence would be much better written with the *gerund* or verb form used as a noun:

> Using a collimating lens throws the *xy* plane to infinity

or with *we* as the subject:

> We used a collimating lens to project the *xy* plane to infinity.

A participle can similarly dangle from the end of a sentence:

> At some load, tearing begins, increasing the rate of rise of the integral.

We won't discuss whether tearing in a plate can influence an integral on paper, but will confine ourselves to the participle, *increasing*. What does it modify in this sentence? Nothing, really. Sentences like this should be rephrased

> At some load, tearing begins and increases the rate of rise of the integral

or

> At some load, tearing begins; this increases the rate of rise of the integral.

Most gerunds used in technical writing are also *-ing* forms of the verb. Like participles, these can dangle from the beginning of a sentence:

> By adopting this point of view, the focus of the analysis is placed on the measurement system.

As before, the phrase that includes the gerund must relate to the subject of the sentence. This sentence should be rewritten

> By adopting this point of view, we focus the analysis on the measurement system.

Alternatively, the gerund itself can be used as the subject of the sentence:

> Adopting this point of view focuses the analysis on the measurement system.

Sometimes it helps to move the gerund phrase to the end of the sentence, instead of locating it at the beginning. That is, change

> By using several approaches, the normal stress was calculated

to

> The normal stress was calculated using several approaches.

Although not especially elegant, the second construction is at least defensible if we regard the last phrase, *using several approaches*, as an adverbial modifier of the verb, *was calculated*.

Subordinate clauses, too, can dangle from the beginning of a sentence. Take pains to ensure that the clause refers to the subject of the sentence, not some modifier of the subject. For example,

> When he was a small boy, Einstein's father sent his son to the gymnasium

might make you wonder how a small boy managed to have a son. The sentence should be rewritten

> When Einstein was a small boy, his father sent him to the gymnasium

or

> When he was a small boy, Einstein was sent to the gymnasium.

Some experts consider an infinitive dangling when it does not refer to the subject, but I argue that such infinitives may often be considered *sentence modifiers* (see **adverb**). For example,

> To maintain saturation, the pressure must be reduced

seems clear and need not be rewritten as

> To maintain saturation, we must reduce the pressure,

though this form is also correct.

Here, without comment, are two dangling prepositional phrases that caused a great deal of mirth at a meeting I attended:

> In an agate bowl, with agate balls, grind for 30 minutes.

See also **misplaced modifier; only; this**.

dash. Sometimes called the *1-em dash* or the *long dash*, this is often useful when you want to highlight some word or phrase. Often you put a parenthetical expression between dashes, because—though important—the expression is grammatically parenthetical.

> Nixon—the only president ever forced out of office —is now being rehabilitated as an elder statesman.

Even though the sentence stands without the parenthetical phrase, the dashes make certain that you won't miss the point. Use dashes in this way, instead of **commas**, when you want to highlight a nonrestrictive phrase or clause.

On a typewriter, the dash is usually faked with two hyphens, and I think it looks better with a space on each side. (In typesetting spaces are not used.) But watch out! Some word processors may get fooled and end a line right in the middle of your dash. To prevent this, I use two superscripted underline characters.

data. Educated usage still dictates that *data* is a plural noun. The singular, *datum*, is almost always passed over in favor of *data point*, which is a peculiar construction unless *data* is singular. And what about **agenda**? Don't ask; see **Ciardi's law**.

database. This word has become so common that it may be considered one word. See also **hyphen**.

date. U.S. usage has it that we write *January 30, 1941*, not *January 30th*, although *the thirtieth of January* is OK. We also do not write *January the thirtieth*, with the definite article.

In military or government style, write *30 January 1941*, or *30 Jan 41*, without commas.

February, incidentally, has two *r*'s, one in the middle and one at the end. It is not spelled *Febyuary* and should not be pronounced that way either.

de-, un-. These, especially *de-*, are too often used to invent some of the worst jargon: *de-mythologize, de-traumatize*, for example. There is also a difference—*de-* connotes removal or moving away from, while *un-* implies negating. After you have *decalcified* something, it is *uncalcified.*

dead words. Orwell said, "If it is possible to cut a word out, always cut it out." Today we might *cut* the two instances of *out* from his sentence, but they are not exactly instances of *dead* or meaningless words. Dead words are phrases like *it is seen that, it is well known that, it is found that, recall that*, and *note that* and its partner *it should be noted that*. These words have no real meaning. When they are used to excess, they do not do the job they are supposed to do: to highlight points you want to make strongly. The best way to highlight some point is to do it with your writing or your punctuation. For example, instead of writing *it should be pointed out that* all the time, find the sentence you really want to stand out, polish it, and give it a paragraph all its own. That probably makes it stand out better than a handful of dead words and saves typesetting besides. See, however, **note that** for a discussion of when to use this phrase.

Other dead words are words like **reason** and **purpose** when they are part of a phrase like *reasons of* and *purposes of*. See also **it; vague words, vogue words**.

debrief. A barbarism. To *brief* someone is to give him information, to inform him. To *debrief* would therefore be to remove the information from his head. In fact, when you *debrief* someone, he is really *briefing* you. *Interview* is a far better word than the ugly *debrief*.

deceased. A lawyer's word for *dead*. See also **euphemism; sacrificed**.

decimal point. See **naked decimal point**.

degree. In the International System or **SI units**, the *degree Kelvin* no longer exists. It has been replaced with the SI base unit, the **kelvin**. The degree symbol and the word *degree* should not be used with the kelvin. The *degree Celsius* is still widely used in publications that adhere to the metric system, whereas the *degree Fahrenheit* is not regarded as a metric unit. The *degree Centigrade* is obsolete; to convert from degrees Centigrade to degrees Celsius, multiply by 1. In addition, I am pleased to report that *degree sahntigrade*, like the *sahntimeter*, has all but vanished, partly because the degree Centigrade has been replaced by the degree Celsius.

Degrees, minutes, and seconds of arc may also be used with the metric system, though the *radian* is the SI unit of angle.

Degree is sometimes incorrectly abbreviated *deg* (with no period); the International Congress on Weights and Measures requires the symbol °: $T = 25°C$, the angle u was 45°. There should be no space between the number and the degree symbol. If you write *degree Celsius* or *degree Fahrenheit* fully, be sure to capitalize *Celsius* and *Fahrenheit*, but not *degree*. For consistency, degrees of latitude and longitude should also be written with the degree symbol.

depending on. Some writers use *depending on* as a sentence modifier. See **adverb**.

deprecate, depreciate. To *deprecate* is to express disapproval; to *depreciate* something is to reduce it in value or to represent it as having a lower value than its true value. You can *depreciate* the importance of someone's theory without *deprecating* the theory itself. Probably because of the similarity to *appreciate*, depreciate is fast becoming a synonym for *deprecate*. The distinction, however, is important, for soon we may have no word for depreciate, in the sense of its original meaning, to reduce in value.

descriptive abstract. An **abstract** that is written like a table of contents, but put into sentence form, is unaccountably called a *descriptive abstract*. Such abstracts are characterized by phrases like *is presented, is discussed*, and *is described*. They tell almost nothing about a paper, indeed, no more than a table of contents tells about a book. Instead of writing descriptive abstracts, make your abstracts informative summaries of your paper. More under **abstract**.

despite the fact that. Even though.

determine. This verb can occasionally be ambiguous, since it means both *to cause* and *to ascertain.* If you write

> Millikan determined the charge-to-mass ratio of the electron,

it is clear that you do not mean that he caused it to have its present value, but, rather, that he ascertained its value. On the other hand,

> They determined the mass of the payload to be 30 kilograms

could mean either that they measured the mass of the payload or that they decided that it was to be 30 kilograms. In the second example, *determined* should be changed and the sentence rewritten, for example, as

> They decided that the payload would be 30 kilograms.

dictionary. The dictionary is not infallible and sometimes sets arbitrary standards, but it is still an invaluable tool for settling questions of spelling, syllabification, and meaning. Use a dictionary and follow its advice unless you have a very good reason not to.

In the U.S., the standard that many editors use is *Webster's Third New International Dictionary of the English Language.* Among desk dictionaries, *Webster's New Collegiate Dictionary* and *The American Heritage Dictionary of the English Language* are considered among the most authoritative. I like them both, but I especially like the *American Heritage*'s usage panel, which gives a clear idea what a cross section of professional writers considers acceptable. Other good desk dictionaries are *The Random House College Dictionary* and *Webster's New World Dictionary of the American Language*. Smaller dictionaries, like pocket dictionaries, may not give enough information.

difference. Difference *in* temperature or *of* temperature? The purist will argue for *of*, and it is perhaps slightly more elegant, but *in* is idiomatically correct.

different. To, from, or than? Easy. *To* is British, *from* is American, and *than* is what Americans really say when you leave them alone.

discovered missing. You can't be discovered missing; if you were discovered, you wouldn't be missing. Likewise, you can't turn up missing.

discrete. Primarily a scientific term meaning *distinct* or *made of distinct parts*, as in *discrete layers*. Not spelled *discreet*, which has the same root but means *prudent*.

disinformation. Derived from a Russian word (what else? ask my conservative friends), this word is really not a bit different from *misinformation*, but it has come to mean *deliberate* misinformation, and we are probably stuck with another inelegant word.

disinterested. This word has so long been used incorrectly for *uninterested* that its more correct meaning, *having no financial interest*, has all but disappeared.

dissociate. Not disAssociate.

double negative. A purely artificial prohibition and an example of what generations of pedantic teachers can do to a language. As we all know, *I don't got none* is considered incorrect because the first negative negates the second. All right. I'll let them have it—even though, to a great many native speakers, this construction sounds more emphatic than *I don't have any*. (Certain other languages, presumably no less logical than English, use double negatives acceptably; possibly the redundancy of a double negative has been found to increase the likelihood of understanding.)

Flesch points out, though, that another kind of double negative is quite common in spoken English.

> Not today, we don't.

That is a double negative; it is perfectly clear; and it means the same thing as

> Today, we don't.

A third kind of double negative is rather stylish. There is a shade of difference between being *able* to do something and being *not unable*. This kind of construction is, however, harder to understand than the more simple statement and should be used only when the shade of difference is important. I had to stop and read twice a sentence telling me that a court had *refused to overturn a law prohibiting* something or the other. In plain English, the court had *upheld* the law.

-down, -out. These are replacing or supplementing *-up*. You watch to see how things *shake out*; prisoners (who always used to be *locked up*) are now subjected to a *lockdown*; wars *wind down* (not up). No wonder people have so much trouble understanding us!

downside risk. Is there an *upside risk*? No, but *downside* and *upside* are becoming firm as clumsy synonyms for *disadvantage* and *advantage*.

downsize. Reduce.

> We will have to downsize some services

should have been

> We will have to reduce [or curtail] some services.

See also **jargon**.

draft. I am constantly revising. I am never satisfied. But I never write a draft or an **outline**. Before I sit down to write a paper, I make sure that I know the material so well that it all comes spilling out when I sit down to write. A draft is a waste of time if it implies that you are going to rewrite it, as opposed to revising it. This is especially so today, when the modern word processor makes it so easy to reorganize paragraphs without Scotch tape and scissors.

Some experts recommend waiting at least a full week between the first or second versions of a paper and the final revision, and I find the practice helpful. The farther you can get from your paper, the less familiar you will be with the words and the arguments, and the more likely to spot errors or inconsistencies. See also **procrastination; wastebasket**.

dry run. You almost never hear anyone say *rehearsal* any more; today, you have a *dry run* before you make your final presentation to the boss. The term was originally military slang for firing small arms without ammunition. Avoid it in formal writing; use *rehearsal* or perhaps *practice session*.

due to. There used to be a prejudice against using *due to*, as in

> Humidifying inside air increases costs due to the energy required to evaporate the water.

Some sources still argue that *because of* is correct here because *due to* introduces an adverbial clause. But I see *due to* used in this way so much that I argue that those sources are outdated, and I see nothing wrong with that sentence. On the other hand,

> The noise falls due to the bandwidth being decreased

is just plain clumsy and should be rewritten

> The noise falls because the bandwidth is reduced.

Due to the fact that is another clumsy synonym for *because*.

duplicate publications. Duplicate publications, it seems to me, are rarely justifiable; neither is a series of papers that report only very slight progress from one paper to the next. An editorial board on which I serve has discussed the problem of duplicate publications several times. We have concluded that duplicate or very similar publications can be justified primarily when the audiences are likely to be completely different or when one of the papers has a very limited audience. If the publications are planned as duplicates, each should refer to the other with a statement that the papers are essentially the same; otherwise, the second should refer to the first with such a statement.

duration. In electrical engineering, this is the term to be used in place of *width*, when you are referring to time interval. That is, use *duration* or *pulse duration* in place of *pulse width*, and *full duration at half maximum* (FDHM) in place of *full width at half maximum* (FWHM). This is something of a victory for the **term-ites**, who argue that width should refer only to geometric quantities.

E

e.g. The abbreviation for the Latin phrase, *exempli gratia*, for the sake of example. In text, I prefer the English phrase, *for example*, though I might accept *e.g.* in a footnote or a **reference**, where we tend to be more cryptic. See also **foreign words and phrases; i.e.**

each and every, each individual. Each.

-ee. It was witty when someone first used *-ee* as the opposite of *-er* or *-or*, as in the legal terms *mortgagor* and *mortgagee* or *employer* and *employee* (now commonly *employe*). But now it has become hackneyed. If you give money to someone, he may be the *donee*, but is he the *givee*? If you sponsor a PhD student, is he the *sponsee*? I suggest a moratorium on neologisms that use this suffix.

effect. Often part of a wasted phrase, as in

> These are plans that will have the effect of positioning the company as a major supplier.

This sentence should be shortened to

> These are plans that will position the company as a major supplier.

See also **affect, effect; dead words**.

effectual. Supposedly very slightly different from *effective*: Effectual is a bit stronger and implies clearly or decisively effective. Still, I will stick with *effective*.

effectuate. No different from *bring about* or *effect*.

ellipsis. In grammar, *ellipsis* means *leaving out a word or phrase that is not necessary for meaning*. For example, Perrin cites

> When [you are] in Rome, do as the Romans do

as an example of ellipsis; the bracketed words are not part of the original quotation but are understood. In technical writing, many **dangling modifiers** are the result of incorrect ellipsis. For example,

> When using a lens with spherical surfaces, rays that pass through the lens at different radii will cross the axis at different points

begins with a dangling participle. If you wanted to retain this style, you would have to complete the *elliptical clause* that begins the sentence:

> When you are using a lens with spherical surfaces, rays that pass through the lens at different radii will cross the axis at different points.

The three dots (sometimes three asterisks) that indicate where you have omitted part of a quotation are also known as *ellipses* (the plural of *ellipsis*). I use four dots when the ellipsis ends a sentence; the fourth is the period found at the end of every sentence. Ellipses should not be used for any purpose besides showing omitted material.

empirical. Does not mean *experimental* or *by trial and error*, but rather *based on fact or observation*. Thus, a theory can be empirical. When, for example, you develop an equation that is not based on theory or on principles, but which you use merely because it fits your data, do not use *empirical* to contrast with *theoretically based*.

employ. Unless you mean *give a job to*, use the shorter verb *use*.

encryption. A relatively new word that has come into prominence because of the use of computers in banking and finance, not to mention spying. Encryption implies a *secret code* and is therefore different from *encoding* (Morse code is code, but it is not secret). *Encipher* is an older word that is synonymous with encrypt.

end product, end result. Why not leave out the word *end*, especially with *result*? *Result* does not need qualification unless you mean to distinguish from *intermediate result* or *preliminary result*.

engage. Usually a **dead word**.

> In our laboratory we have been engaged in measuring attenuation for many years

does not need *engaged in*.

English units. Scientific papers should be written in **SI units**. In the U.S., however, English units—foot, pound, quart—are still common in commerce and in everyday life. If you think it is necessary to use English units in your paper, put them into a secondary position after the correct SI unit. For example, write 5 cm (2 in), with the SI unit first. Label the axes of a graph with the SI units in the primary positions, along the bottom and left edges. If you want to add English or other non-SI units, place them along the top and right edges, preferably in a smaller typeface.

When writing English units, follow the rules for SI units: leave a space between a number and the unit, do not end the abbreviated form of the unit with a period or make it plural with an *s*, and do not mix complete spellings and abbreviations. Similarly, use superscripts, raised dots, and slashes as in SI units: write ft^2, not sq. ft.; ft/s, not fps (for feet per second); and ft^3/min, not cfm (for cubic feet per minute). Some authors abbreviate *inch* as in. when there is a possible ambiguity because of the similarity with the preposition *in*, but the practice seems inconsistent to me, and I avoid it. See also **CGS units**.

ensure, insure. There used to be a difference, and I think it is worth preserving. I use *ensure* for *make certain* or *guarantee*, and *insure* for *take out an insurance policy*. Thus, I *insure* my car against theft, but I lock the doors to *ensure* that it will not be stolen.

enter into. Enter, when you mean a room, but *enter into an agreement* is idiomatically correct.

entitled. *The paper by Swartzlander entitled "Digital Optical Arithmetic"* may be arguably correct, but it is awfully formal and would be better as *The paper, "Digital Optical Arithmetic," by Swartzlander*. (I always thought that *entitled* meant *to have a right to*.)

equally as. As in *equally as great as*. Usually, the *equally* should be dropped:

> The pressure in one arm was equally as high as that in the other

should be

The pressure in one arm was as high as that in the other

or

The pressures in the two arms were equally high.

equations. Very short equations may be used as nouns in sentences:

Points $r < a$ are in the core of the fiber.
The line extends from $x = a$ to $x = b$.

Longer equations are given lines of their own and usually numbered along the right margin. These equations are also best treated as parts of sentences and, in most publications, are also punctuated as such:

The lens equation is

$$(1/l') - (1/l) = (1/f'), \tag{23}$$

where f' is the focal length of the lens.

It is incorrect to write

Calculating the focal length of the lens from (22),

followed by a comma or, worse, a colon, and the equation. In this construction, the participle *calculating* is a **dangling modifier**, since it does not properly modify anything in the sentence.

It is poor style, or at least very formal, to write

We arrive at the following expression for the lens equation:

and follow it by the equation on a separate line. Instead, write

We arrive at the expression,

$$(1/l') - (1/l) = (1/f'), \tag{23}$$

for the lens equation.

In this construction, the equation is in **apposition** to *expression* and is therefore set between commas.

Although it is perhaps not wrong, avoid a construction like

From Eq. (3), the sum of the stresses is

$$S_3 = S_2 + S_1. \tag{11}$$

Instead use

Equation (3) shows that the sum of the stresses is

$$S_3 = S_2 + S_1 \tag{11}$$

or

We find from Eq. (3) that

$$S_3 = S_2 + S_1. \tag{11}$$

In none of these examples is there a comma before the equation, since the equation is treated as just another word or phrase.

Some publications ask you to write your equations on one line whenever possible; this saves typesetting costs. In addition, writing equations on one line is often desirable in a singly spaced typescript, where superscripts or subscripts sometimes spill over onto the adjacent lines and reduce legibility. Be sure, though, that you use parentheses when they are necessary to avoid ambiguity. For example, don't write

$$pV = 1/3\,Nmv^2;$$

it could be mistaken for

$$pV = 1/(3Nmv^2).$$

Instead, write

$$pV = (1/3)Nmv^2.$$

More generally, if you precede a complicated denominator with a slash, enclose the denominator in parentheses:

$$\sin\theta = x/(x^2 + y^2).$$

-er, -est. See **comparative, superlative**.

essentially. Much overused, usually as an incorrect synonym for *nearly*. *Essentially* really means *in essence* or *characteristically* and should not be used in a construction like *essentially free of oxygen* if the intended meaning is *nearly*. See also **vague words, vogue words**.

et al. The abbreviation for the Latin phrase, *et alii*, and others. (There is no period after *et*.) I think that *et al.* should be avoided almost entirely: In references, it is polite to include the names of all the authors and not brush off their contributions with a four-letter abbreviation. When there are more than, say, five authors, however, I would accept *et al.* on the grounds of practicality. In text, though, I prefer the English phrases *and others, among others*, or *and her colleagues*.

When I use *et al.*, I punctuate it exactly as I would the English expression *and others*. That is, I would precede it with a **comma** only if it is part of a list that has more than two entries: *Einstein et al.*, but *Kaplan, Eisenstein, et al.* See also **foreign words and phrases**.

etc. The abbreviation for the Latin phrase, *et cetera*, and others. Not, incidentally, pronounced *eKcetera*. In references, I might use *etc.* sparingly, but in text I prefer the English phrases *and so on, and so forth*, or *and the like*. *Etc.* should not be used after *included* or *including*, since these imply that you are not listing the entire set. See also **comprise; foreign words and phrases; include**.

euphemism. A more-or-less neutral word that is substituted for a word that is considered distasteful or too strong, like *gender* for *sex* or *defense* for *war*. Don't use euphemisms; say what you mean.

We also need the antonym *dysphemism* for using complex phrases or unusual words like *chief executive officer* when we mean *president* and *fenestration* when we mean *windows*. Don't use dysphemisms, either.

every day, everyday. *Everyday* is an adjective, as in *an everyday occurrence*. But when *every day* appears with *every* modifying the noun *day*, write the phrase as two words:

We back up our hard disk every day.

If you are in doubt, see whether you could substitute *every other day* for *every day*. If it makes sense, spell *every day* as two words. See also **everybody, everyone**.

everybody, everyone. These are singular and supposedly take singular pronouns: *Everyone go to his seat.* The trouble with this rule is the masculine pronoun and the fact that everyone says *their seats.* Still, I try to use the singular pronouns with *everybody* and *everyone*. See also **collective noun; he or she**.

Sometimes, *every one* is a phrase with *every* modifying the noun *one*:

> Every one of them got relief from the medication, so the study was ended early.

If you are in doubt, see whether you could substitute *every last one* for *every one*. If it makes sense, spell *every one* as two words. See also **every day, everyday**.

evidence. Is it a transitive verb, as in

> The survey evidenced the dependence of cancer on lifestyle?

Maybe it is, but *showed* or even *exhibited* are better choices. *Evidence*, as a verb, is stilted and sounds like jargon.

Evidence, incidentally, comes in two kinds, *anecdotal evidence* and what I call *hard evidence*. Anecdotal evidence is evidence based on stories or anecdotes told by individual observers, whether trained or not. Anecdotal evidence should not be regarded as proof: it may be wrong. Rather, anecdotal evidence, when it is convincing, gives you a hunch or an idea what to look for. Only after you have performed some sort of controlled study or experiment are you entitled to regard your evidence as hard evidence or proof. Uncritical acceptance of anecdotal evidence is at the core of most beliefs in pseudoscience.

exclamation point. Strunk and White say never to use an exclamation point at the end of a complete sentence, but rather to reserve the exclamation point for **sentence fragments**. Well! That may be a bit extreme, but their point is a good one: Use your writing, not an artificial punctuation mark, to make your points strongly. For example, don't write

> The material became superconducting at 4 K!

Instead, you might set this sentence off in a paragraph of its own as a way of highlighting it. (See also **it**.) Similarly, however exciting I thought supernovas were, I would not have titled an otherwise sober article

> Supernova 1987A!

but would have found some other way of making the title interesting.

From time to time, however, you will find a use for an exclamation point at the end of a complete sentence. You might sound peevish if you responded to a review with the words

> This reviewer certainly has strong opinions.

But an exclamation point might show that you are good-humored about the review:

> This reviewer certainly has strong opinions!

Although they are unusual in technical writing, you could also end a command or other imperative with an exclamation point.

exists. Overused as a synonym for *is*. For example,

> There exists a theoretical foundation for these measurements

would be less clumsy as

> There is a theoretical foundation for these measurements,

though you could write

> A theoretical foundation for these measurements exists

if you preferred. Do not, however, write

> A theoretical foundation exists for these measurements.

Here, the prepositional phrase *for these measurements* is a **misplaced modifier**. See also **it; there**.

expect. Not a synonym for *anticipate* or *suspect.*

> We expect that the results will be difficult to interpret

should have been

> We anticipate that the results will be difficult to interpret.

experience. Often a **dead word**.

> Experience shows that such delays are constant over a long period

adds nothing to the flat statement

> Such delays are constant over a long period.

F

fabricate. Usage has it that you *fabricate* integrated circuits, for example. Usually, however, *fabricate* is not a bit different from *make* or *build* and should be replaced by a shorter word.

fact. A fact is different from an assumption. A fact is something that has been verified. If something is not true, it is not a fact. Therefore, you cannot say

> His facts were not true.

What you mean is that they are not facts. Similarly, *true fact* and *actual fact*, like the bankers' *free gift*, are redundant.

factoid. An interesting neologism, derived from *fact* + *-oid*, the Greek suffix for *like*. A factoid is a statement that is designed to look like a fact, but is either wrong or irrelevant. Suppose, for example, that you state there is no evidence that a certain diet cures arthritis. A common response to this kind of statement is

> Physicians are not trained in nutrition.

This is a factoid: whether or not it is true (and I suspect it is not), it is irrelevant. It neither refutes your statement nor provides evidence to support the diet's claims. A factoid is an ignoramus's method for diverting you to an irrelevant topic.

factor of. Write a *factor of 5*, not *a factor of five*. See **numbers in sentence**.

false elegance. Using long or pretentious words where common words exist. Many of the entries in this book, for example, using *utilization* for *use*, *in view of the fact that* for *because*, and *therein* for *in it* are examples of false elegance.

> Much advantage accrues to this method from the availability of fluorescent dyes

is a falsely elegant way to say

> Fluorescent dyes have proved valuable to us.

Other writers achieve (if that is the right word) false elegance by *elegant variation*, or changing words to avoid repeating a word within the same sentence or paragraph. Indeed, bad writers often go to great lengths to avoid repetition. Why? You don't get variety, say, by switching to *fabricate* just because you have just used *build* or switching to *employ* because you have just used *use*. Use the word that sounds most natural to you, and don't worry about getting variety by an artificial device.

> We calculated the mean and standard deviation; the average was 17.0.

Average and *mean* are synonyms here, but why change terminology in mid-sentence? Say

> We calculated the mean and standard deviation; the mean was 17.0.

See also **circumlocution; gobbledygook; wordiness**.

far-distant. A redundant expression that means *very distant*. Instead of

> Computer simulations eliminated a far-distant cloud of comets,

write

> Computer simulations eliminated a very distant cloud of comets.

farther, further. These are variations of the same word, and we could do with just one of them. The custom is, however, to use *farther* for distances and *further* for all other cases. *Further* seems somewhat clumsy for *additional*, as in a *further problem*. *Further to your inquiry* is a **stock phrase** that should be avoided entirely.

fast rate, slow rate. Rates are not fast or slow; things are. Rates are *high* or *low*.

faster time. Shorter time.

fat cat's law. Cover your assets.

feedback. This word has a legitimate place in electrical engineering and certain other disciplines; otherwise it is a **jargon** word. I do not want your *feedback*, however much I may value your *opinion*.

The verb is *feed back*, two words:

> The power is fed back into the cavity.

See also **vague words, vogue words.**

female, male. When referring to people, these are best left as adjectives and *woman, man* used as nouns. Likewise, avoid euphemisms like *fair sex* or *distaff side*. Many women today would be properly insulted if the meaning of distaff had not been forgotten.

few number. This should be replaced with *small number* or *few*, depending on context. You can't have a few number of things, only a few things or a small number. Similarly, *few in number* is redundant; leave out *in number*.

Another strange construction is *x number of things*. Just as you hear *three things*, you would expect to hear *x things* or *an unknown number of things*. Still, *x number* has become colloquial and is perhaps acceptable in speech if not in writing.

fewer, less. Use *fewer* for countable things, *less* for quantity: *fewer* marbles, *less* water. Many people do not make this distinction in speech any more, but it should be preserved in writing. Instead of

> It contains one less amino acid than Protropin,

write

> It contains one amino acid fewer than Protropin.

ff. The abbreviation for *and the following pages*. I would restrict using this abbreviation to footnotes and **references**. See also **foreign words and phrases**.

figure. Use figures fairly liberally to illustrate your points, especially in popular articles or articles for specialists in other fields. Generally, a figure is a **graph**, a photograph, or a line drawing. In a technical paper, use a photograph of your apparatus, for example, only if it really adds something to your exposition. Most of the time, scientists know what scientific apparatus looks like, and a line drawing or a schematic drawing serves better. In a popular article, however, there is nothing like a good color photograph to elicit the desired *gee whiz* from your readers.

figure caption. Sometimes a terse caption is in order, especially if it makes a point. For example, in an article on cosmetic defects in glass, there is nothing wrong with a drawing or a photograph with the caption

> Optical quality is in the eye of the beholder.

In general, though, make your captions informative. If you present a **graph** with a great deal of information in it, don't make the casual reader dig through the paper to find out about the graph. Rather, make the caption complete enough that the reader can understand the figure from the caption alone.

finalize. Another yucky *-ize* word. Use *complete* or some other synonym. See **-ize, -ization**.

finite. This means *not infinite*, but it also means *not infinitesimal.* It therefore means any number that is neither zero nor infinity. It is not the same as *nonzero*, since infinity is not zero.

first, second, third. These are perfectly good adverbs and do not need **-ly**.

first measurement, first report. How do you know yours is the first report? Qualify such statements by saying as far as you know, yours is the first report.

first person. It is not known by the author—oops! I do not know why there is such a prejudice in the scientific community against the use of the first person, especially *I*. I suspect that it has to do with the idea that science and the scientific method exist outside their practitioners. What I do know, however, is that the avoidance of *I* and *we* makes for clumsy, unclear writing. Many times, for example, I have read introductions to technical papers and figured out only with difficulty when the authors stopped reviewing the literature and started telling what they had done. Why? Because every sentence had the same structure: This was done, that was done. When I review a manuscript with this defect, I always advise the authors to take credit for what they have done.

In my own writing, I oscillate between *I* and *we* (when I am the only author). Thus, I write *I* when I mean *I*, and *we* when I mean *the reader and I*. For example,

> Using a sewing needle, I carefully punched a 25 micrometer pinhole into a piece of brass shim stock,

not

> Using a sewing needle, a 25 micrometer pinhole was punched into a piece of brass shim stock.

But, on the other hand,

> We therefore see that any thermal source may be made coherent if its angular subtense is made small enough.

In the last sentence, *we* means you and I: the reader and the writer.

Using *one* is another device for avoiding the first person. At its worst, it is pompous, as when the critic writes in *The New York Times Book Review*, *One is tempted to infer that* .... When writing a technical paper (or anything else that goes out over your name), speak for yourself, not for some indefinite *one* and certainly not for everyone else. Just for yourself. In addition, the more personal pronouns you write, the easier your manuscript will be to read.

The prejudice against *I* extends to the objective case, *me*.

> A similar result was obtained by Young and myself

should properly be

> A similar result was obtained by Young and me

or, better,

> Young and I obtained a similar result.

Why not *myself*? Because *myself* is a *reflexive pronoun*, that is, one that refers to the subject. It is a synonym for *me* only in a sentence like

> I got the result myself,

where it is used to intensify the subject, *I*.

Avoid using *I* when there are two or more authors, even though only one of you actually wrote the paper. In particular, I thought it was inappropriate when one author identified himself as the originator of the project and then referred throughout the paper to *I* and *Roberta and I*. Not only is *I* the wrong pronoun to use when there is more than one author, but its use also puts your co-author into an unacceptably subordinate position.

Flesch's readability index. In the 1940's, Rudolf Flesch developed two scales for readability. He called one of these his *reading ease score*. It has reappeared once or twice since, but, as far as I know, Flesch's was the first. Flesch's index is a number that is calculated from the average sentence length and the average density of polysyllabic words. Modern computer programs that purport to analyze your writing probably use Flesch's readability index or some variant of it. Flesch gives a nomogram for calculating your reading ease score, but, since this is a technical book, I can let on that you may also use the formula,

$$RE = 207 - 1.0 \cdot SL - 0.85 \cdot SY$$

when I have rounded Flesch's numbers to an appropriate number of significant figures. RE is the reading ease score, SL is the average sentence length, and SY is the average number of syllables per hundred words. Your reading ease score is a number that ranges from 0 (practically unreadable) to 100 (very easy). According to Flesch, comic books score between 90 and 100, "quality" magazines around 65, "academic" writing between 30 and 50, and "scientific" writing between 0 and 30.

Flesch tells you to pick samples of your writing according to a formula and not to try to pick "typical" samples. For example, he recommends three to five samples of an article and 25 to 30 of a book. Begin each sample at the beginning of a paragraph, but exclude the introductory paragraphs, since they are often not a representative sample of the work as a whole. First, count the number of words in an integral number of sentences, stopping as close as possible to 100 words in each sample. Count numbers, dates, contractions, equations, and compound words as single words. (You will have to use your judgement to decide what is a compound word; for example, *lock-in amplifier* is obviously two words, not three, but what about *core-diameter measurements*?) Count independent clauses separated by *and* or *but* as one sentence, but count those separated by semicolons or colons as two sentences. Calculate the average sentence length *SL*.

Next count the number of syllables in each sample. This is not as tedious as it sounds: Since you have already counted the words in each sample, just count the second and succeeding syllables in words that have two or more syllables. Add the number to the number of words in the sample. Count the syllables the way you pronounce the words, and count numbers like 14 as two syllables, *four* and *teen*. If you have equations, ignore them in your syllable count, but for each equation add one word from the sentence following your sample and count its syllables instead. (Don't forget the hyphenated words that you have already counted as single words; these should be regarded as single, polysyllabic words here, too.) Calculate the number of syllables per 100 words.

Now calculate your reading ease score *RE* according to the formula. A well-written scientific paper can easily have a reading ease score of 60 or 70, provided that the sentences are short enough. You can't get much higher than that, because scientific writing necessarily uses a lot of long words. But you can compensate for these long words by concentrating on keeping your sentences short. See also **sentence length**.

flip. A word you use when someone is being witty but you do not agree with what he is saying.

fluorescence. Not flOUrescence. *Florescence*, incidentally, is a different word and means *flowering*. Not long ago, I saw the phrase, *the fluorescence of the human mind*. Unless the human mind glows in the dark, the author meant *florescence*.

flora, fauna. Singular nouns. The plurals, *floras* and *faunas*, are almost never used. See also **bacteria**.

folks. It is all right to be colloquial, but some people are trying to be too folksy these days. I draw the line at *the folks at Ballistic Missile Development*, a phrase I really heard someone use.

foolish consistency. Emerson's often misquoted aphorism is

> A foolish consistency is the hobgoblin of little minds.

If something is technically correct but sounds bad, change it. This applies mostly, I think, to sticking unreasonably to the **principle of parallelism**, insisting on the present **subjunctive**, and refusing to use a **split infinitive** or end a sentence with a **preposition**.

footnote. If it is important, say it in the text; a footnote (other than a **reference**) always looks like an afterthought. An exception, of course, is a *note added in proof*, which is usually identified as such and describes a result too recent to have been put into the body of the paper.

for. This preposition is often used inappropriately, sometimes where *of* or *on* would be better. For example,

> Measurements for the same quantity in different
> laboratories often give different results

should really have been

> Measurements of the same quantity in different
> laboratories often give different results.

You do not, after all, make measurements *for* a quantity; you make measurements *for* a purpose or a sponsor.

for all intents and purposes. Not *for all intensive purposes. This* **cliche** is, in any case, a **redundant expression**. Find some other way to say it.

foreign words and phrases. In some circles, it is de rigeur to stick a Latin or French phrase into your repartée. A propos of this, why? To show what a maven you are? There is almost always a suitable English locution that means the same thing; why not use it? Then there will be almost no chance of your being misunderstood. See also **cf.; e.g.; etc.; i.e.; ibid.**, all of which can easily be replaced by a perfectly good English equivalent. (Now that I have said that, I can't think of a good, crisp synonym for *status quo* or *ad hominem*. Oh, well.)

If you decide to use a foreign expression, it is probably best to italicize it, unless it has really become English, like, perhaps, *status quo* or *ad hoc*. Still, if you use a Latin phrase as a modifier, you may get into trouble if you do not italicize it. For example, I did a double take at the sentence

> In hormonal treatments or in vitro fertilization, it developed in response to pressure from the infertile.

What is *vitro fertilization* and how do you get in it? The phrase should have been hyphenated, *in-vitro fertilization*, or the Latin words written in italics.

Foreign phrases that virtually no one has ever seen before should not be used at all:

> The introduction of domesticated animals, *sensu stricto*, is a topic of much debate.

former. See **last, latter; parentheses**.

found to be. This phrase is almost always found to be unnecessary.

from. Write *from 100 to 200*, or *100–200*, but not the hybrid, *from 100–200*.

From is also commonly used in the context

> From this table, it is clear that simple lenses improve with decreasing focal length

or

> From Eq. (8), the ion scatter rate is mostly the result of the background gas.

Better style is to write

> The table shows that simple lenses improve with decreasing focal length

or

> Equation (8) shows that the ion scatter rate is mostly the result of the background gas.

(*Equation* is spelled fully in the second instance because it begins a sentence.) See also **between; it**.

A lot of **circumlocutions** begin with *from*.

fun. *Fun* used to be a noun. But, since we never use it with the article *a* or *the* (it was fun; he is fun), it began to be mistaken for an adjective. Besides *a fun thing*, which could be defended as using an attributive noun, you will now hear *It was so-o-o-o fun!* Like it or not, *fun* has become an adjective. Well, why not? It's sort of fun following these things.

further. See **farther, further**.

future plans, past history. All plans are for the future, and all history is past. With the minor exception that you have present plans and are thinking of your future plans, both phrases are redundant.

fuzzy logic. Although this is a technical term in computing, I use it to describe sentences that don't quite make sense, even though they may be grammatically correct. For example,

> The economy, in the form of the gross national product, has not kept pace.

The gross national product is not a form of the economy; it is one way of describing or quantifying one aspect of the economy. The author meant

> The economy, as measured by the gross national product, has not kept pace.

Similarly, in

> The physical characteristics of these systems are similar to those of PEG-dextran systems, except for the cost,

cost is not a physical characteristic. The sentence should have been

> The physical characteristics of these systems are similar to those of PEG-dextran systems, but they cost more.

Finally,

> Longsworth reached a temperature of 79 K in a two-stage refrigerator

makes me hope they got poor Longsworth out before it was too late.

Sentences like these are one reason for beginning a paper early enough to allow a week or so before you prepare the final **draft**. The most poorly prepared papers I read are those written by authors who are always running late against a deadline. See also **procrastination**.

G

game plan. Plan.

gedankenexperiment. Use the English *thought experiment.* See also **foreign words and phrases**.

gender, sex. *Gender* is a grammatical term and describes whether a noun is masculine, feminine, or neuter. It is not synonymous with *sex*, but, since some wag invented the phrase *gender gap*, we are stuck with *gender* as a **euphemism** for *sex*.

Much of the outcry about "sexist language" is the result of the confusion between gender and sex. See **he or she; man, -man**.

general public. Public.

general trend. Trend.

generic name. See **trade name**.

gerund. The *-ing* form of a verb, when it is used as a noun. The prescriptivist insists on its being modified by the possessive case, as is *being modified* in this sentence. That is, *it being modified* might be considered technically incorrect. *Being modified* is not a participle here and does not modify *it*, as in the similar construction, *the part being examined*, for example.

How can you distinguish between a participle and a gerund? Try leaving out one of the words. For example, you could not say *The prescriptivist insists on it by the possessive case* . . . , but you could say *The prescriptivist insists on being modified by the possessive case*. . , at least grammatically. Therefore, *being modified* functions as a noun, and must be preceded, or modified, with the possessive case, *its*.

Sometimes, however, the possessive case sounds funny, and many prefer to stick with it only when the modifier is a pronoun. A small point, at best. Write whatever sounds best to you. See also **dangling modifier**.

getting started. I have confessed to a great deal of **procrastination**, but, even so, it is eventually necessary to sit down and write. How to begin?

Some authors simply start with references to past work:

> High pass filtering in optical processors was described by Birch [1] and by Swing [2]. A zero was found in the image of an edge, and it was proposed to use this to measure linewidth. The result was applied to microscopy and an error function was calculated.

They may spend several sentences or even paragraphs before they tell you about their own paper. In this example, the last sentence refers to the author's work, but you have to figure that out for yourself.

If you want to start out with the background or the history of your problem, it would be better to begin with a narrative history:

> When a high-pass filter is used in an optical processor, sharp edges stand out brightly in the image. Several years ago, Birch [1] and later Swing [2] noticed that the bright edge contained a sharp zero of intensity located precisely at the geometrical image of the edge. They proposed using this zero as an aid in measuring linewidths in microscopy .... In this paper, I will take a fresh look at the problem and show that the width of the filter should be approximately one-fourth the width of the transform plane.

Blicq suggests that you begin with the phrase *I want to tell you that* ... and finish the sentence with what he calls your Main Message:

> I want to tell you that high-pass filtering can be used to measure linewidths in microscopy.

Then write your paper (or your **introduction**), and go back and erase *I want to tell you that*. This technique forces you to identify what you most want to tell, as well as to identify your audience. Identifying the audience is important, and some writers overlook this step, especially when writing memos or reports.

Alternatively, you can begin by writing *The purpose of this paper is to* . . . . This phrase may be deleted or let stand, though you may get a bit tired of it after a while. (If you delete it, you will have to rewrite the opening sentence.)

Tichy suggests two dozen "beginnings." These refer not to the first sentence but to the introductory paragraph or paragraphs. Most are not germane to formal papers or reports, though they can be used effectively in mass market or trade magazine articles, memos, and occasionally journal articles.

The most useful beginnings in technical writing are probably opening with the purpose of the paper,

> The purpose of this paper is to describe our efforts to measure and characterize scratches on polished glass surfaces,

with a summary of the main ideas or results,

> Recently, we sponsored a conference on optical surface quality standards. Approximately 50 scientists attended the day-and-a-half conference and listened to a dozen invited papers and a panel discussion. During the panel discussion and in private talks with my colleagues and me, the participants agreed on the following points: . . . In addition, they agreed that no single standard would satisfy every need,

with a statement of the scope of the paper or of the problem it addresses,

> Many gas lasers have adjustable mirrors and may be operated in a single mode or, to gain power, in several transverse modes. We have examined the effect of multimode operation on the coherence of the laser beam,

or with the background for the research,

> The pinhole camera or *camera obscura* was one of the earliest optical instruments. Alhazen used the pinhole camera in about 1000 to study a solar eclipse [1], but its invention is often credited to della Porta in about 1600 [2]. In addition, Mach thinks that

> Leonardo may have used the pinhole camera for studies in perspective [3]. Despite this antiquity, the pinhole camera boasts several advantages over more sophisticated optics.

It is, incidentally, perfectly acceptable, sometimes desirable, to state your conclusions in the introduction, so that your readers will have an idea where the paper is leading them.

The best ways to start a formal paper are probably to discuss the background, purpose, scope, problem, or ideas in the paper. Sometimes your introduction or opening paragraphs may combine backgroud with purpose: begin with a short paragraph of background, and then state the purpose or scope of the paper. The paragraphs on background should be written in a narrative or descriptive style and should not, like the first example in this entry, be simply a list of references or a table of contents in sentence form.

Other beginnings are really devices for getting started. Many are appropriate for magazine articles or **conference proceedings**, but only rarely for journal articles. For example, you could begin with a joke,

> Since learning that I was to present my paper at 5:30 on the evening of the Hallowe'en party, I have had the recurring fantasy that there would be only one person left in the audience. I would give my presentation and conclude, "Thank you very much for listening," and she would reply, "Quite all right. I am the next speaker,"

a question,

> Is radiation loss really a factor in determining the efficiency of a greenhouse?

a quotation, an example, or even an outrageous statement designed to get your audience's attention,

> Only three people have ever understood the scratch-and-dig standard. One is dead, one is insane, and I, the third, have forgotten.

(This quotation, incidentally, was cribbed from Lord Palmerston and acknowledged in a footnote.) If you use one of these devices, though, use it with care, and make sure that your opening is both appropriate and written in a style that is consistent with the rest of the paper. The opening quotation or example, for instance, should be relevant to the topic of the paper, not simply a gratuitous attention grabber. See also **organization**.

given. Much overused, often as a synonym for *because*. For example,

> Given that the instruments were not optimized,
> precision never approached the theoretical value.

Some authorities would consider *given* used in this way a **dangling modifier**. In any case, the sentence would be better rewritten as

> Because the instruments were not optimized, precision
> never approached the theoretical value.

See **adverb**.

glitch. Seems to me to be a perfectly good word, which need not be put into **quotation marks** or otherwise apologized for.

go that route. Much too colloquial to use in writing.

> 132 nations have pledged not to go the nuclear
> weapons route

should have been rewritten

> 132 nations have pledged not to develop nuclear
> weapons.

gobbledygook. Derived from *gobble*, this means using *fenestration* for *windows* or *adjudicator* for *judge*, and otherwise using a long word where a short one will do.

> Refrain from counting on anything until it is finalized

is gobbledygook for

> Don't count your chickens before they are hatched.

Other examples of gobbledygook are *progress accounting document* for *report, wildlife conservation officer* for *game warden, adversely impacting upon* for *harming*, *personnel station* for *desk*, *enjoy superior longevity* for *live longer*, and *imputation of data* for, well, *fudging*. See also **redundant expressions; vague words, vogue words**.

good. An adjective, not an adverb. Something works **well**, not *good*.

gotten. I once heard this disparaged as *a vulgar Americanism*. But my dictionary says it's OK, and my colleagues vote for it with their feet (so to speak).

graduated. *Was graduated*, as in

> He was graduated from Rochester in 1962,

is archaic and a bit pompous. Leave out *was*.

graph. Let us call the labels put onto the x and y axes the *labels*. Under the label for the x axis, there may be a *caption*, or a title for the graph. The little marks that denote the scales are called *ticks* or *hatches* (not *tick marks* and certainly not *tic marks*; a tic is a nervous twitch).

The labels should be written in the form *Force, newtons* or *Force (newtons)*. The comma is probably better than the parentheses, since the parentheses can have mathematical meaning that might be confusing. Especially avoid the **slash**: *Force/newtons* looks too much like a fraction.

Similarly, don't try to get too much into the label. I have seen things like *Force/F/N* for *Force, F, in newtons*. If you have to get all that stuff into the label, do it discreetly: *Force (F), newtons*.

On the other hand, be sure to put enough information into the labels. In particular, state the quantity, not just its units. That is, do not label an axis *Kilojoules*, but, rather, *Energy per pulse, kilojoules*.

Be consistent within the same graph and, when possible, from graph to graph in the same paper. For example, don't abbreviate the label along one axis and spell fully the label along the other. When you abbreviate the name of a unit, make sure that the abbreviations are correct SI abbreviations (see **SI units**).

Capital and lowercase lettering usually looks better than all capitals for printing the caption and the labels. Use the same rules you use for capitalizing titles. That is, either capitalize the first letter only, or capitalize the first letter of each word that is not an article or a preposition. My habit is to capitalize the first letter only. See **capitalization**.

When you submit a graph (or any other figure) to a journal, it will usually be reduced to a one-column width. The lettering should be large enough to remain at least 2 mm high when the figure is reduced. For a typical journal, this comes to about 3 to 4 percent of the height of the figure. Alternatively, the lettering in *points* (seventy-seconds of an inch) is about 4 times the height of the graph in inches. As a rule, Gothic or *sans serif* lettering looks best in a graph that is reduced to one or even two columns. The serifs, or decorations, on the more complicated letters give a cluttered appearance to typical graphs, which have very small lettering.

Tufte advises keeping graphs as simple as possible. For example, he points out that a rectangular grid superimposed on a graph can often hide the data or make an overall pattern or trend hard to detect. So can fancy graphing techniques, like drawing bar graphs with three-dimensional bars. Tufte calls these embellishments *chartjunk*. To avoid chartjunk, plot only what is necessary: the axes and their labels, short hatches instead of a grid, and the data. Do not put too many curves or symbols, or confusing amounts of data, on a single graph, and keep text that lies within the boundary of the graph as short as possible. For example, if you have a lot of different symbols on the graph, either leave plenty of room to display them, or else display them in the caption, outside the graph itself. See also **slides, viewgraphs**.

ground rules. Rules.

H

harebrained. The Strategic Defense Initiative is not *hair brained*; it's *harebrained.*

hard copy. A take-off on *hardware* but not yet spelled as one word. A *hard copy* of a document or a picture is a rendition on film or paper, as opposed to magnetic disk or tape.

hard-wired. Permanently connected. It was not correct to write

> Millan took pictures from the sidelines using a hard-wired remote flash,

since the remote was not permanently connected to Millan's camera. The author meant *wired*, but tried too hard to distinguish between a flash that was wired to the camera and one that was triggered remotely by the flash on the camera.

has the property that it is. I once saw this lengthy synonym for *is* in a manuscript.

hatched. See **crosshatched**, which is sometimes used incorrectly for *hatched.*

having. Constructions like

> We use a gas laser having improved performance

are common, but I think it is better to write

> We use a gas laser that has improved performance

or

> We use a gas laser with improved performance.

Whenever you are tempted to use a present participle, see whether you can replace it with a prepositional phrase or a *that* clause. As Bremner points out, if you don't use participles, you can't dangle them. See **dangling modifier**.

Another problem with *having* is that it is a present participle, but is often used with a verb in the past tense:

> A transmitter having a frequency of 100 MHz was used in the experiment

has a problem with **tense**; it should be rewritten

> A transmitter that had a frequency of 100 MHz was used in the experiment.

he/she, s/he. See **he or she**.

he or she. It is unfortunate that English has no genderless pronoun—or, rather, that the pronoun *he*, which is often used for both sexes, is the same as the masculine pronoun. Like the genderless suffix *-man*, *he* can be misunderstood to imply maleness in an era when women have every right to compete in any profession. (Pity the poor French, who have to say *Madame le docteur*, with both the masculine ending and the masculine article.)

But what to do about it? Clumsy locutions like *he or she* or, worse, *s/he* and *he/she* do not solve the problem: they point it out by adding barbarisms to your writing. I often alternate between *he* and *she*, as in this book, but that, too, sometimes seems forced.

One solution is to write in the plural: Instead of

> Each student has his own computer,

write

> All students have their own computers.

(A permissivist, I suppose, would accept

> Each student has their own computer,

but I wonder what he or she (ha!) would say about

> Someone looking into such a mirror would see themself as others do.)

You are not necessarily "sexist" if you use *he* to mean *he or she*. Unfortunately, many readers will either assume you are or take umbrage at your using *he* in that way. Therefore, when all else fails, perhaps it is time to bite the bullet and use *he or she* for your generic pronoun. See also **gender, sex; man, -man; man-made**.

head up. In *She heads up the lab*, delete *up*.

healthful, healthy. These are not synonyms. Eating *healthful* foods will make you *healthy*. The foods themselves are not healthy unless they are growing right outside your door.

heavy, light. The lighter of two light things is not half as light; it is half as heavy. But it would be better to say it weighs half as much.

height. *Length, width, breadth*, and *depth* notwithstanding, *height* does not end in *h* and is not pronounced *heighTH*.

herein, therein. Pompous words that make your writing very formal. Instead of

> A solution to this problem is presented herein,

write

> We propose a solution to this problem in this paper.

heretofore. Previously or since then.

his/her. See **he or she**.

hitherto. Previously.

homogeneous. HomoGEEneeous, not homAHgenous.

hopefully. Someone must have thought that *hopefully* was the opposite of *regrettably*. Regrettably, it is not, and it just dangles when used at the beginning of a sentence in place of *I hope*. Most authorities do not accept *hopefully* as a *sentence modifier* (see **adverb**) and prefer that it be used in its original sense, *in a hopeful manner*, as in

> "Did the Mets win today?" Matt asked hopefully.

Hopefully can also be used evasively:

> Hopefully, we will complete the project on time

may well mean

> I hope we will complete the project on time, but I doubt it.

however. This *conjunctive adverb* may also be used as a sort of flavoring word; however, when it is used to join two clauses, it should be preceded by a semicolon, not a comma. Stylists, however, consider it better not to begin a sentence with *however*, but rather to set it off with commas in the middle of the sentence. However, there is nothing really wrong with capitalizing *however* and beginning a sentence with it. However you use it, absence of a comma after it indicates a different meaning. All four uses of *however* are demonstrated in the four preceding sentences.

human resources department. Personnel.

hyphen. Technical writing sometimes looks as if some *mad dasher* had taken a hyphen shaker and sprinkled it over the page. This is a result of using many compound nouns as modifiers; it is to some extent unavoidable because technical writing has to be very precise. When do you use hyphens and when not?

Many publications require hyphenating all compound modifiers, like *gas-laser beam, chemical-vapor deposition, optical-fiber waveguide*, or *10-cm radius*. (See **numbers in sentence; SI units** for rules on hyphenating with numbers.) When writing for these publications, therefore, consistently hyphenate such constructions.

The purpose of the hyphen is to link the two words in the compound modifier, which may be two nouns, an adjective and a noun, or even three or four words. (See **adjective; inanimate possessive**.) Usually, these modifiers are hyphenated only when they directly precede the noun they modify. That is, we would hyphenate a phrase like *host-specific parasite*, but would not hyphenate the same phrase as a *predicate adjective*, as in

> These lice are host specific.

A hyphen could indicate a considerable difference in meaning. For example, *popular science writer* could be understood to mean either *a writer of popular science* or *a popular writer of science*. Insertion of a hyphen, *popular-science writer*, makes clear that the meaning is *a writer of popular science*. Frequently, however, the meaning is clear from context or from common usage. I do not think it necessary to hyphenate *gas laser beam*, because no one will think you meant some kind of gas beam. Similarly, I see no need to hyphenate phrases like *more powerful rocket* or *high inten-*

sity lamp, since these are not ambiguous. The trend nowadays is away from hyphens, and using hyphens sparingly makes your writing look more colloquial and less like **jargon**. Therefore, when writing reports or conference proceedings articles, where you are not constrained by a publication's **style manual**, consider making your writing more informal by omitting hyphens where they are not necessary for clarity. If, however, you use a lot of compound modifiers, especially three-word modifiers, I recommend hyphenating throughout. See also **adverbs**.

It is important to hyphenate correctly. For example, *nuclear-directed energy weapon* should have been *nuclear directed-energy weapon*, since it is not a *nuclear-directed weapon* but rather a certain kind of *directed-energy weapon*.

Certain expressions, nevertheless, cry out for hyphenation. Odd juxtapositions of words, like *on-off switch* or *nitrogen-rich mixture*, should be hyphenated. Indeed, any expression, like *lock-in amplifier* or *in-situ measurement*, that uses a preposition in a funny place or a phrase that involves numbers, like *three-part harmonies* or *first-magnitude star*, should be hyphenated, no matter what hyphenation style you are using in general. I was taken aback by the sentence

> Most devices today are designed using titanium in diffusion.

The author meant *in-diffusion*; in other words, the titanium is diffused into the device. Writing *in diffusion* as if it were a prepositional phrase like *in solution* left an ambiguity.

Do not, however, hyphenate a *phrasal verb*, that is, a verb like *log on, back up*, or *lock in*, which involves a particle similar to a preposition. On the other hand, hyphenate adjective forms of such verbs, or spell them as one word: *back-up copy, coronary bypass operation*. Here usage, familiarity, and, perhaps, taste should be factors in your decision whether to hyphenate the compound or combine it into a single word.

Modifiers that involve prepositional phrases, like *signal-to-noise ratio* or *lack-of-fit method* should also be hyphenated. You would not want people to think that you lacked a fit method.

In addition, I hyphenate whenever I use a prefix that doubles a vowel or creates a possible ambiguity like *ie*, as in *electro-optic*, *anti-evolution*, or *mono-unsaturated*. Similarly, I use a hyphen for a construction like *non-Gaussian*, because of the capital *G*. With these exceptions, a hyphen is not necessary with most prefixes and suffixes: *nonnegative, quasithermal, uniaxial* are the preferred forms. You might, however, want to write *pseudo-intellectual* because of the *oi*, and hyphenate things like *un-ionized* and *per-iodate* to avoid possible confusion with other words. *Co-author* perhaps deserves a hyphen to break up the three consecutive vowels.

Some style manuals arbitrarily call for a hyphen after a longish prefix like *quasi-* or *pseudo-*. (These, incidentally, are *prefixes*, not words, and must be connected to a word. See also **prefix, suffix**.) Some authors like to hyphenate a compound like *x-axis* or *J-integral* when it involves a single letter, even when the compound stands alone and is not used as a modifier. What is important is to be consistent throughout your manuscript. Do not, for example, hyphenate *J-integral* but not *x axis*. Similarly, if you hyphenate *x-axis*, then write *xy-plane*, not *x-y plane*.

Do not hyphenate *speed of light* or *state of the art*, except when they are used as modifiers; these phrases are unambiguous in sentences like

> The speed of light is now a defined quantity.

Likewise, do not hyphenate compound nouns except when they are used as modifiers: *a shock tube* but *a shock-tube refrigerator*. Finally, do not hyphenate adverbs that end in **-ly** in constructions like *optically active material*; since the adverb is obviously linked to the modifier, there can be no ambiguity.

Eventually, compounds become so common that we begin to write them as one word. In my own field, many authors write *farfield diffraction, laserbeam*, and *blackbody*, not to mention *searchlight*. Check the latest edition of a good **dictionary** for the most recent usage. With newer compounds, you may have to decide for yourself; one clue to when a compound has become a single word is the pronunciation. When a compound is accented on the first syllable, it is time to start spelling it as one word. Until that time, hyphenate it only when it is used as a modifier.

I

I. If you did something, take credit for it. Avoid *the undersigned, the author, the writer*. See also **first person**.

I am writing. As Flesch says, of course you are writing! Otherwise I wouldn't be receiving your letter. Avoid this and many other **stock phrases**.

i.e. The abbreviation for the Latin phrase, *id est*, **that is**. Except perhaps in a reference, I prefer the English expression. Sometimes used incorrectly to mean *for example*. See also **foreign words and phrases**.

I personally. This is a funny way of saying *As for me, I* . . . or *I, for one,* . . . . How else can you think something? Impersonally or corporately? Avoid writing or saying *I personally*.

ibid. The abbreviation for the Latin word, *ibidem*, in the same place. It is used in **references** or **footnotes** to repeat a reference to the book, chapter, or article most recently cited: *ibid., p. 17*. In scientific or report writing, *ibid.* is comparatively rare, since references are frequently not as specific as they are in scholarly writing. In addition, scientific reference style permits a reference number to be repeated in the text when the reference is to the same work. Consequently, there is less need for *ibid.*, and it is more common to use something like *Ref. 6, chap. 3* when you want to refer to a specific part of an earlier reference. *Op. cit.*, in the work cited, and *loc. cit.*, in the place cited, are almost the same as *ibid.*, except that they may be used to refer to any earlier reference: *Gould, op. cit., p. 308*, or *Gould, loc. cit., note 3*.

identical. Not a synonym for *the same*. In

> The voltage was measured on the identical device,

the authors meant

> The voltage was measured on the same device.

They should have used identical only if they meant

The voltage was measured on an identical device.

See also **same identical**.

if. If you use an *if* clause to begin a sentence, then you do not necessarily need to follow it with the adverb *then*. If the sentence is short, then it may seem clumsy and the adverb out of place. In any case, if you use *if . . . then* repeatedly, then your writing may look more formal or stilted than you intend.

If and when should almost always be shortened to *if*. *If* is conditional, whereas *when* is not. Therefore, you must mean one or the other. Most of the time, *if* is appropriate, not *when*.

If is sometimes used after an expression of doubt, where **whether** would be preferable.

impact. This is not really a verb, except in the sense of impacted wisdom teeth. It should be replaced by *influence* or *affect*, as should the clumsy expression *have an impact on*.

As for *negative impact*, I agree with the person who said that a negative impact occurs when time is reversed, a heavy weight rises from a physicist's foot, and the physicist says *!hcuO*. The better expression is *adverse effect*.

imperative. Urgent or necessary. *Imperative* should not be used as a synonym for *important*; it is a much stronger word. Nor should *imperative* be qualified with a word like **absolutely**; something is imperative or it is not. See **absolute words**.

imply, infer. I may *imply* something; you may *infer* something completely different. You can't ask

What are you inferring?

unless you mean

What inferences are you drawing from the evidence?

The correct question is usually

What are you implying?

in, into. You come *into* contact with something; later you may remain *in* contact with it. A quarterback throwing *in* a crowd is surrounded by defenders; a quarterback throwing *into* a crowd has a receiver surrounded by defenders. In short, *in* is often misused as a synonym for *into*. Try substituting *into* in place of *in*. If it makes sense, *into* is the right word.

Likewise, do not use *on* when you mean *onto*. Make the same test: if you can substitute *onto* for *on*, then *onto* is the word you want.

in and of itself. Itself; inherently.

in many cases. While editing the manuscript for this book, Aidan Kelly pointed out my tendency to use phrases like *in many cases*, where in many cases the shorter *often* would serve. Kelly says

in some cases = sometimes,
in many cases = often,

and

in most cases = usually.

If it were not for my dislike for **foreign words and phrases**, I would be tempted to write *mea culpa*. Instead, I will just note what a pleasure it is to learn from an expert.

in nature. Usually redundant, as in

> Most low power lasers are ophthalmic in nature,

which is just as clear as

> Most low power lasers are ophthalmic.

in order to. Usually equivalent to *to*, as are *so as to, in an effort to*, and *in such a way as to*. But occasionally the words *in order* are necessary to imply intention.

> It was necessary to measure the thermal losses

means that we had to measure the thermal losses, whereas

> It was necessary in order to measure the thermal losses

means that something was necessary if we were to measure the thermal losses.

in regard to. Not *in regards to.* Usually this is an imprecise phrase that can be replaced by *regarding* or *concerning*.

in shape. Usually superfluous.

The cavity is cylindrical in shape

or

The cavity is cylindrically shaped

should be more compactly written

The cavity is cylindrical.

in size. Another superfluous or redundant expression. You need not write

The antenna was small in size,

since

The antenna was small

is adequate. You might also want to change

The antenna was one meter in size,

to the more specific

The antenna was one meter in diameter.

See also **vague words, vogue words**.

in vacuo. In a vacuum. See also **foreign words and phrases; in vitro, in vivo**.

in view of the fact that. Because.

in vitro, in vivo. These are Latin phrases, but there are no acceptable English synonyms. I would therefore use them, but italicize them. When you use such phrases as modifiers, hyphenate them if you cannot italicize them. See also **foreign words and phrases; hyphen; in vacuo**.

inanimate possessive. Pedants prohibit this construction because, though you can own a chair, the chair cannot own a leg. Even Flesch says that an inanimate possessive sounds funny and should be replaced by a prepositional phrase. Sometimes this is so, but how about *a stone's throw* or *a dollar's worth*?

Many technical writers avoid the inanimate possessive by simply using the noun as an **adjective**, as in *the condenser Airy disk diameter*; this would be better as *the condenser's Airy disk diameter*, although *the diameter of the condenser's Airy disk* is arguably the best form. The important thing, it seems to me, is to be clear, and this usually means avoiding a string of nouns. The possessive breaks up the string as effectively as a prepositional phrase.

inasmuch as. Because.

include. Do not use **etc.** or *and so on* with *includes*, because *includes* already implies that you are not listing the entire category. When you are listing the entire category, do not use *includes* or *including* but, rather, *consists of, contains*, or possibly *comprises*. See also **comprise**.

Leave out *included* from a sentence like

> Included in this category are shawms, cornets, and sackbuts,

and write instead

> In this category are shawms, cornets, and sackbuts,

or, better, reverse the sentence order:

> This category includes shawms, cornets, and sackbuts.

indicate. This means *show, point out* (both of which are shorter). It does not mean *say*, so it is usually not correct to say *he indicated*.

indigenous. This means *native* and almost always refers to a country or a region. It is really stretching a point to write

> The microscope is an indigenous part of many measurement systems.

The author probably meant *inherent* or *integral part*.

infinite. An **absolute word** that should not be modified with adverbs such as *nearly* or *practically*. A quantity that is so large that it may be regarded as infinite is still not *nearly infinite*.

information for contributors. See **manuscript preparation**.

informative abstract. A short summary or synopsis of your paper. An informative abstract is not written as a table of contents, but rather summarizes your paper and tells its conclusions. With few exceptions, abstracts to scientific papers should be informative abstracts. See **abstract**.

input, output. Like **feedback**, these words have a legitimate place in certain engineering disciplines. I do not, however, want your *input*, however much I may value your *suggestions*.

Input and *output* should be used as nouns, not verbs, except in the very specialized usage,

INPUT C$,

within a computer program. A computer program may also *output* data. With these exceptions, the verbs should be *put in* and *put out* or, preferably, some synonym. For example,

A femtosecond dye laser outputs similar pulses

sounds better as

A femtosecond dye laser puts out similar pulses,

but best as

A femtosecond dye laser produces similar pulses.

inside of, outside of. Why the superfluous preposition *of*? Cut it. *Inside of an hour*, however, is idiomatically correct.

instant in time. Instant.

instead of, rather than. You will hear that the **principle of parallelism** requires constructions like

He theorized rather than experimented.

Frequently, especially with the past tense, these sound forced; most native speakers would say

He theorized rather than experimenting.

Still, I would be careful not to mix parts of speech:

> He was a theorist, instead of experimenting in the laboratory

should be rewritten

> He was a theorist, instead of an experimenter.

I can see no difference between *instead of* and *rather than*, though sometimes one sounds better than the other. Each phrase is a unit, and it is not significant that *of* is a preposition and *than* is a conjunction. *Instead of*, for instance, does not have to be followed by a noun or a gerund just because *of* is a preposition.

interconnect. This is a legitimate word, but it means *connecting three or more things* and should not be used as a synonym for *connect.*

interface with. You may interface with other people, but I don't. At least not in the lab.

interpretative. *Interpretive* is better, but *explanatory* is best.

into. See **in**.

introduction. An introduction should be more than a listing of references to earlier work. Use it to make your paper more interesting or more accessible to readers who are not as thoroughly immersed in your subject as you are. For example, use it to give the background to your research: what has already been done, why the subject is important, what is new about your work, and what it contributes. One formula that works is to begin, *The purpose of this paper is to* . . . or *In this paper, I* . . . .

Additionally, you can outline the subject as descriptively as possible or make your introduction a summary of the paper to follow. State your hypothesis and, perhaps, your **conclusions**. Your paper will get the widest exposure if you write it in such a way that it can be read even by someone who is not fully conversant with your subject. One way to appeal to such readers is to include as much background as you can in your introduction and to write it as simply—as nontechnically, you might say—as you can. See also **abstract; getting started; organization**.

intuit. A barbarism, but an old one.

intuitively obvious. Maybe not redundant, but just a shade short. See also **trivial**.

-ion, -ity. Often signs of writing that is too abstract. In technical writing, abstractions are often hard to avoid, but you can cut down on constructions like *the stabilization of the laser* and replace them with *stabilizing the laser*, which sounds less stilted and also saves a couple of syllables.

irregardless. There is no such word as *irregardless*, which probably results from the confusion between the synonyms *irrespective* and *regardless*.

is dependent on. Depends on.

is, is. In the sentence,

> The problem is, is that the world has too many guns and not enough butter,

the sentiment is unarguable, but what about the syntax? Perhaps the speaker meant to say

> What the problem is, is that the world has too many guns and not enough butter,

but that is not what came out. This *is* with a grace note is unfortunately becoming common in speech; please don't write it. Write instead

> The problem is that the world has too many guns and not enough butter.

is of. Usually, *has* will do at least as well.

> The solution is of the form $y = mx + b$

is a little more clumsy than

> The solution has the form $y = mx + b$.

Similarly,

> Patients are of short stature with stocky trunks but thin limbs

should be rephrased as

> Patients are short and have stocky trunks but thin limbs.

Watch out also for *is indicative of*, which is usually not a bit different from *shows* and is indicative of **wordiness**.

it. Prescriptivists argue that you can't begin a sentence with *it*, as in

> It is the high coherence of the laser that gives it its remarkable properties,

because the subject *it* has no antecedent. This sentence, however, is not one whit different from a sentence that begins *There is*. . ., and I can see nothing wrong with it. Indeed, it is stronger than

> The high coherence of the laser gives it its remarkable properties.

Judicious use of *It is* . . . can make your writing more powerful.

Most of the time, however, the word *it* introduces a completely meaningless or superfluous phrase like *It should be noted that* . . . or *It can be shown that* . . . . These are wasted or **dead words** and are used partly to avoid the **first person**. Such phrases add no emphasis to your writing, especially when they show up all the time. I would blue-pencil such phrases into oblivion, even if the sentence has to be completely rewritten. A particularly egregious example I recall is

> It should be noted that it is not necessary that this test be automated,

which translates nicely to

> This test need not be automated.

Similarly, in introductions to technical papers, we often see

> It has been shown that the connectors can cause an error in the power measurements,

perhaps followed by a reference. Sometimes it is difficult to decide whether the authors are referring to earlier work or telling about their own paper, which they call *the present work*. It would be much clearer to write

We have shown that connectors can cause an error in the power measurements

and refer to the present work as *this paper*.

Phrases like *It is clear from Fig. 3 that . . .* often make me wonder, Clear to whom? This cliche has become a joke. Better to write *Figure 3 shows that . . . .* Likewise, *It is reasonable that* means *I think that*, and *It is conceivable that* really means *I would like to believe that*.

Some other wordy phrases and their replacements:

it is probable that . . . can usually be replaced with *probably;*

it follows that . . . can usually be omitted or replaced with a judiciously placed *therefore*;

and

it was desired to . . . usually means *we wanted to . . . .*

For example,

Since *a* approaches 1, it follows that *b* approaches 0

does not need the phrase, *it follows that*. On the other hand,

It follows that the flux density is proportional to the radius

is more concise as

The flux density is therefore proportional to the radius.

Finally, *it* is often used superfluously in a sentence like

I would like to make it clear that the new superconductors have many useful properties.

Here you could omit *it*, not for some theoretical reason, but because it contributes nothing to the meaning or the intensity of the sentence:

I would like to make clear that the new superconductors have many useful properties.

See also **circumlocutions; wordiness**.

its, it's. The possessive case of *it* is *its*, no apostrophe. (There is no apostrophe in *his, hers, yours, ours*, or *theirs*, either, but these do not seem to cause as much confusion.) *It's*, with the apostrophe, is the contraction for *it is* and should never be used in place of *its*.

Despite what the purists say, I can see nothing wrong with *it is me* or *that's him*. These have become idiomatically correct, though I would usually avoid such constructions in formal writing. (I confess, however, that *this is her* and *this is him* sound dreadful to my ears.)

-ize, -ization. You may not be able to create something from nothing, but you can create verbs from nouns by adding *-ize*. I actually saw *de-Sandinistaizing the army* once in an otherwise fairly literate journal. I cannot think of a worse way to say *ridding the army of Sandinistas*. Likewise, *deprojectization* is a barbarism for *canceling a project*. Use of the suffix *-ize* usually makes for clumsy writing that looks full of **jargon**.

We are probably stuck with some *-ize* words. *Prioritize*, ugly as it is, is here to stay, even though *rank, order*, or *set priorities* mean roughly the same thing. Other words, like *stigmatize* or *dramatize*, are perfectly acceptable English, as are technical words like *crystallize, hydrolyze*, or *sterilize*.

Sometimes a barbarism like *connectorize* will find a niche for itself. (This means not *connect*, but *put a connector onto the end of*, and is usually used in the sense of a *connectorized cable*.) Still, I think we should not needlessly make up verbs or use jargon words where ordinary synonyms exist.

J

jargon. This is not necessarily a pejorative term: To the linguist *jargon* is *the specialized vocabulary of a group or profession*. Technical terms are, therefore, jargon. When a jargon word escapes and is taken over by the general population, it becomes *slang*; eventually it may either become *standard English* or disappear. *Clone* is an example of a scientific word that has a precise meaning, yet has become slang for *an identical copy*. *Paranoid, schizophrenic*, and other words from psychiatry have also become slang words with somewhat less-precise meanings than their technical meanings.

Jargon is sometimes used to imply unintelligibility. One of the most well-known and most egregious examples of unintelligible jargon is the airline steward's *extinguish all smoking materials* for *put out your cigarettes*. Using long words or expressions where short ones will do also makes your writing look like jargon. So does inventing abbreviations where none already exist. Your writing will be clearer and more accessible to nonspecialists and specialists alike if you refrain from using technical words except where they are necessary for precision or clarity.

Use jargon when it is necessary, but don't be a *jargonaut*!

Johnson's law. No expense will be spared in trying to find a free lunch; when it is finally found, every effort will be made to find someone else to pay for it.

justification. See **right justified**.

K

kelvin. The unit of temperature, abbreviated K, and no longer °K. That is, write 273 K, not 273°K or 273 °K.

In speech, say *300 kelvins*, just like *300 kilograms*. See also **degree; SI units**.

keyboarding. Typing, but usually at a computer or a word processor. An old, mechanical word processor is known as a *typewriter*.

kilodollar. Although the dollar is not an SI unit, the prescriptivist might argue for the form 10 k$. Well, it just isn't done that way. A dollar sign is always written before numerals, and, in addition, it has become customary to write a capital K: $10K.

kilometer. The preferred pronunciation is *KILuhmeeter*, not *kilAH-muhter*. See also **SI units**.

kind, sort, type. The plural of *one kind of plant* is *two kinds of plant*, not *two kinds of plants*. There is no need to make the object of the preposition plural. Similarly, *those kind of things*, which you hear mostly in speech, should properly be *that kind of thing* or *those kinds of things*, depending on meaning. The use of the plural *those* to modify the singular *kind* shows that the speaker got confused by the upcoming plural *things*. See also **singular or plural**.

kind of, sort of. *A kind of plant*, not *a kind of a plant*. But an expression like *kind of a long way* for *a somewhat long way* is idiomatically correct (without the article *a*).

kudos. Singular, like hubris. You do not give *a kudo* for something; you give *kudos*. Pronounced *KYOO-dos*.

L

last, latter. Constructions that use *the latter* are usually harder to understand than alternatives. If you insist, use *the latter* for two things but *the last* for three or more.

> The dye may be dissolved in water, ethanol, or glycol, but only the latter has sufficiently high viscosity

makes you go back to see which compound is the *last* (not the *latter*). It would be clearer to repeat *glycol* and say

> The dye may be dissolved in water, ethanol, or glycol, but only the glycol has sufficiently high viscosity

or rewrite the sentence to state it another way. There is no good reason to avoid repeating *glycol*. See **false elegance**.

The following sentence is real, and I would rewrite it if only I knew what it meant:

> This latter annulus is slightly larger than the image of the clear annulus so that it completely covers the latter.

See also **parentheses**.

lastly. Last. There is no need to attach **-ly** to last, which is already a perfectly good adverb.

Latin abbreviations. Like **e.g.; et al.; etc.; i.e.; ibid.** I can see no reason to use any of these, except perhaps in **references**. Likewise, I can see no reason to use **foreign words and phrases** when perfectly good English synonyms exist.

latter. See **last, latter**.

laundry list. List.

laws. From time to time in this book, I quote certain "laws." These are **Baumeister's law; Buckley's law; Ciardi's law; fat cat's law; Johnson's law; Mrs. Young's law; Pascal's law; technocrat's law**. In addition, some rules laid down by Orwell are detailed under **barbarisms; dead words**, and an aphorism by Emerson under **foolish consistency**.

lay. See **lie, lay**.

left-hand side, right-hand side. These may be shortened to *left side, right side.*

legitimate. This is a legitimate adjective, but the verb should be *legitimize. Legitimatize* is a barbarism.

lend, loan. *Lend* is the verb, and *loan* is the noun. The bank does not *loan* you money; it *lends* you money (and then only if you have money to start with).

lens. Not lense.

less. See **fewer, less**.

letter. See **business letter**.

level. In

> The field was measured at several stress levels,

delete *level*:

> The field was measured at several stresses.

See also **dead words; vague words, vogue words**.

liaison. Lee-AY-zon, not LAY-uh-zon.

lie, lay. If I were designing a language for the humor of it, I would have made the past tense of *lie* the same as the present tense of *lay*. Even native speakers can't keep these two verbs straight. *Lie* is intransitive: I *lie* (myself) down. *Lay*, on the other hand, is transitive and requires a direct object: I *lay* something down on the table. You cannot *lay* down for a nap; you *lie* down for a nap.

The *principal parts* of these verbs are *lie, lay, lain*; and *lay, laid, laid*: I *lie* down now; *I lay* down yesterday; I *have lain* down for an hour now. I *lay* the book on the table now; I *laid* the book on the table yesterday; I *have laid* the book on the table already. Sometimes, to be sure, I have to recite the principal parts to myself before setting pen to paper.

The participles are *lying* and *laying*: I am *lying* down for a nap, but I am *laying* a book down on the table.

Next time you hear someone say, "I am going to lay down for a while," bite your tongue.

ike. A preposition, not a conjunction. In formal writing, *like* should not be used to introduce a subordinate clause. See also **as, like**.

I can see nothing really wrong with using *like* as a synonym for *such as*, though sometimes *such as* sounds better. For example,

> Several imaging systems, like the pinhole camera, use neither lenses nor mirrors

just doesn't sound as good as

> Several imaging systems, such as the pinhole camera, use neither lenses nor mirrors.

In this sentence, *such as* is used to introduce an example, and the sentence could have been written

> Several imaging systems, the pinhole camera, for example, use neither lenses nor mirrors.

On the other hand, you cannot use *such as* for a comparison:

> The zone plate, like the pinhole camera, images without using either lenses or mirrors.

How to choose between them? A good rule to follow is, If *such as* fits, use it. Otherwise, use *like*.

ke that. This must be a regionalism.

> It seems like that we ought to be able to do it.

More standard speech would drop *that*. In writing, use *as* or *as if*. See also **as, like**.

ke when. I saw nothing wrong with the sentence,

> It's like when you are riding a bicycle,

when Miss Cahill caught me using it, and I see nothing wrong with this construction today. The *when* clause is nothing but a noun clause, the object of the preposition *like*.

kely. Most commonly used as an adjective:

> The experiment is likely to yield a null result,

but you can also use it as an adverb, in place of probably:

The experiment will likely yield a null result.

linking verb. Certain verbs that describe a state of being or becoming can function as *linking verbs*. These verbs *link* the subject of a sentence with a noun or adjective in the predicate. The mos common linking verbs are *be, become, look, feel, seem, smell, taste* and *sound*. These verbs require a *predicate adjective* where many incorrectly place an adverb. For example, *to feel poor* is to be under the weather (or out of money); *to feel poorly* is to have a poor sense of touch. Likewise, someone *tests normal*, not *normally*; to test normally would be to conduct a test in a norma manner.

list. Lists should obey the **principle of parallelism**. That is, all the items in the list should have the same form: all nouns, all sentences, all participles, whatever. Avoid something like

RESEARCH LABORATORY
1. Gowns must be worn.
2. Gloves must be used.
3. Wash hands.

Here, items 1 and 2 are passive sentences, whereas item 3 is a command. The last item could be changed to

3. Hands must be washed,

but the list would be more forceful if all three items were written as commands:

RESEARCH LABORATORY
1. Wear gown.
2. Wear gloves.
3. Wash hands.

literally. A word many people use when they mean *figuratively*, as in

I literally died laughing.

Really? You are looking very good considering.

litmus test. This test is a definitive test for acidity, so *litmus test* in slang means a definitive test, not just any test. See also **acid test; redundant expressions**.

loc. cit. The abbreviation for the Latin phrase, *loco citato*, in the place cited. See **ibid**.

locution. A word, a phrase, or an expression. Rather formal, but I do not know a good synonym.

long length. I know you can have a *short length*, but a *long length* sounds terrible. You would never say *strong strength*; you would say *great strength*. So say *great length* for the opposite of *short length*.

In this regard, I saw the expression *high strength, long length fiber* in a journal. This would have been much better as *long, high strength fiber* or *long, strong fiber*. Indeed, *a long length of fiber* is just *a long fiber*.

lowercase letters, uppercase letters. Capital letters are sometimes called *uppercase letters*, because printers used to keep them in a case located above the case for the small, or *lowercase*, letters. Usually abbreviated *caps* and *lc* or *l.c.* When a title, for example, is to be changed from capital letters to a combination of lower- and uppercase letters, the editor usually writes *caps and lc*. See **abbreviations, acronyms; capitalization; graph; references; trade names**.

-ly. This suffix, which is used to form the **adverb** from the **adjective**, sometimes seems in danger of disappearing. *Nearly perfect*, for example, has almost entirely been replaced by *near perfect*. Use the correct adverbial form, *nearly*. Similarly, *various-sized couplers* should have been written *variously sized couplers* (or *couplers of various sizes*).

The common failure to use *-ly* can also result in an unwitting ambiguity. For example, in the sentence

> One experiment used a single shielded wire in a harness,

did the author mean *only one shielded wire* or *a singly shielded wire* (a wire with only one shield)? As it turned out, the sentence was correct as written, but could have been made clearer by rewording,

> One experiment used one shielded wire in a harness,

or by adding a comma,

> One experiment used a single, shielded wire in a harness.

The suffix *-ly* is commonly added superfluously to *firstly, secondly*, . . . , and incorrectly to *thusly*. (*Thus* is already an adverb, never an adjective, and it is incorrect to make it *thusly*.)

Similarly, certain verbs called **linking verbs** require the adjective rather than the adverb, but many writers incorrectly use the adverb.

-ly words are never followed by a **hyphen**, because they are obviously adverbs and do not need to be linked to the adjectives that follow them.

M

man, -man. These used to be neuter but nowadays are considered sexist. Changing *man* to *humanity* is fairly simple, but the suffix *-man* is harder to eliminate. *Fire fighter* and *mail carrier* may lack the snap of the originals, but they are relatively harmless. Unfortunately, excessive zeal to change supposedly sexist words often leads to barbarisms like *waitperson* or, worse, *waitron.* I wish merely tampering with the language would solve the problem. See also **gender, sex; he or she**.

man-made. *Synthetic* or *artificial* will sometimes do as a neuter replacement, but *artificial changes in the climate* does not place the problem on our doorstep, where it belongs. *Anthropogenic* has the aroma of jargon. See **gender, sex; he or she; man, -man**.

many. See **as many as; multiple; numerous**.

manuscript preparation. Journal articles, books, and magazine articles are usually *typeset.* That means that someone else, the editor, is responsible for laying out the pages, selecting the typefaces, reducing the figures, and so on. All you submit is a manuscript and figures, and probably check the proofs. The editor does the rest.

Your responsibilities, however, include submitting a neat manuscript, typed or printed with a very good dot matrix printer. The manuscript need not be letter-perfect, but may have a few neat, penciled corrections. If you cannot type some symbols, write them neatly in pencil as well. Identify (also in pencil) handwritten or other unusual symbols, *Greek mu, proportionality sign*, for example.

In addition, study the journal carefully and follow its **style manual** as closely as possible. If you do not have a style manual, get one. Many major scientific publishers, such as the American Institute of Physics, the Institution of Electrical and Electronics Engineers, and the American Psychological Association, publish their own style manuals. In addition, every journal I am familiar with publishes a page or two of "Information for Contributors" at least yearly. Book publishers, too, follow certain rules of style. You will help the editor speed the publication of your article if you follow the publisher's rules of punctuation, abbreviation, and so on.

For example, prepare a separate title page. If the journal capitalizes only the first letter in the title, type your manuscript that way. If the journal prints your name and address directly below the title of your article, prepare your title page accordingly. On the other hand, if the journal prints a sentence, such as *The author is with the National Bureau of Standards*, at the foot of the first column, type such a sentence at the bottom of your title page.

In the body of your paper, follow the journal's style regarding, for example, abbreviations. If the journal uses the forms *Fig. 10* or *Eq. (3)*, type your manuscript that way, whatever your own preferences. Copy editors change these things (and they should!); there is no need to make more work for them.

Trade and popular magazines might not have a style manual, but you may be certain that they follow certain rules, possibly those of the *Chicago Manual of Style*. If you plan to publish in such magazines, make yourself familiar with each publication and see how they operate. For example, do they capitalize Figure, Table, and Equation? Do they abbreviate Eq.? Do they number headings and subheads? When you prepare your manuscript, do it their way.

When you are ready to submit your paper, get some reviews of your own. Your own colleagues may have valuable suggestions, and, if they can't understand something, the odds are that the journal's referees and, ultimately, its readers will not either. In addition, Laurence Passell, writing in *Physics Today*, says that you will do both yourself and the editor a favor if you suggest eight to ten experts you think would be competent referees. Far from being presumptuous, you will at least give the editors a chance to expand their database and increase your chances of receiving a competent and informed review.

A manuscript that will be reproduced directly from your typescript is usually called *camera-ready*. A camera-ready manuscript must be typed with a carbon ribbon or printed with a letter-quality printer, such as an ink-jet printer. Follow the instructions exactly; this time there will be no copy editor to catch your errors. When possible, have an editor or a layout expert, not a typist, prepare the final copy. Work the figures and tables into the manuscript, rather than attach them in a block at the end.

Most **conference proceedings** require you to supply a camera-ready article. Likewise, you need camera-ready material for a thesis or a report, like those of government agencies, that will see widespread distribution. Even a short in-house report should be typed neatly, with attention to consistency.

Many camera-ready reports are typed on special *reduction paper* that is designed to be reduced by 77 percent. The reduced report is printed on 22 × 28 cm paper.

Sometimes, the conference organizers will send you detailed instructions for preparing the manuscript. Although you have more freedom than with a typeset journal, these instructions are their style manual; try not to deviate unnecessarily from the instructions.

Instructions or no instructions, the responsibility for laying out the pages still rests with you. You will have to decide on the placement of figures and tables, for example. For guidelines, look at other papers in the same series. See also **figure; figure caption; graph; table**.

mass, weight. Mass differs from weight: mass is an intrinsic property of matter, whereas weight is a force that results from gravitational attraction. Therefore *weight percent*, for example, should properly be *mass percent*.

In the sciences, it is important to preserve this distinction. The kilogram is the unit of mass, not of weight; a weight is a force, and its units are newtons. Therefore, be sure always to write

> The mass was one kilogram,

not

> The weight was one kilogram.

The weight is the mass times the acceleration due to gravity; in this example, it is 9.8 newtons if the mass was near the surface of the Earth. See also **atomic mass, atomic weight**.

maximum, minimum. The phrases, *a maximum of* and *a minimum of*, can often be replaced with *at most* and *at least*.

may or may not. Use only when you want to emphasize the negative or conditional nature of your sentence. See also **whether**.

may be recognized as. Is.

medium. Singular, as in *a recording medium* or *a publication medium*. The plural could be either *mediums* or *media*. I prefer the English plural, *mediums*, as in

> Radio is still one of the important mediums of communication.

The media, however, means newspapers, magazines, and television. Is it singular or plural? I think plural, since you could write

> Book publishing is one communication medium.

Therefore treat *media*, like **data**, as a plural noun. But when you need a singular noun, use *medium*.

See also **publication medium**.

micron. The obsolete term for *micrometer*. Avoid *micron* and its abbreviation μ. Likewise, use *nanometer* in place of the obsolete *millimicron*. See **SI units**.

militate, mitigate. Often confused. To *militate* is to act for or against something; to *mitigate* is to make less severe or to moderate. *Mitigate* is most often used in the lawyer's sense of *mitigate damages* or *mitigate the situation*.

minimalize. Minimize.

minuscule. Not minIscule.

misplaced modifier. A modifier must be located as close as possible to the word it modifies; if not, the reader may infer the wrong meaning or at least be taken momentarily aback. For example,

> Photons can be used as carriers of information as well as electrons.

Really? Photons can be used as carriers of electrons? No. The sentence should be

> Photons, as well as electrons, can be used as carriers of information.

Another common feature of technical writing is a subordinate clause that follows the verb or the direct object instead of the subject, which it is supposed to modify:

> Bacterial strains are emerging at increasing frequency that are resistant to antibiotics.

A frequency that are resistant to antibiotics? No. You can correct your syntax and save two words by writing

> Bacterial strains resistant to antibiotics are emerging at increasing frequency.

In this version, *resistant to antibiotics* is properly located directly after what it modifies, *bacterial strains*.

An adjectival modifier is similarly misplaced when it follows a verb directly; this is particularly prevalent in abstracts or introductions, where authors frequently use the construction, *A method is developed that...*, as in the sentence,

> A method is developed that allows the temperature to be measured precisely.

Perhaps authors use this construction because they are afraid the verb will be lost at the end of a long, complicated sentence. Indeed it might, but then the sentence should be divided into two or more sentences, not written ungrammatically. In any case, the *that* clause modifies the subject *method* and should follow it directly:

> A method that allows the temperature to be measured precisely is developed.

The error can also be avoided by writing in the **first person**:

> We have developed a method that allows the temperature to be measured precisely.

Here is a sentence that is too long by one dangling modifier:

> The causes of alcoholism are varied and complex, only some of which can be attributed to the patient's will.

The phrase, *of which*, is too far from *causes*, which it is supposed to modify. The sentence should be made two sentences or two independent clauses by replacing *which* with *these*:

> The causes of alcoholism are varied and complex; only some of these can be attributed to the patient's will.

When I read a sentence like

> Expert systems drawing on the capabilities of human experts stored in computerized databases will draw conclusions,

I wonder whether they let the experts out for a walk now and then. The sentence should be rewritten

> Expert systems that draw on the capabilities of human experts will be stored in computerized databases and draw conclusions.

The misplaced modifier can show up anywhere, and there is no simple rule to avoid it. Just make sure that your clauses and phrases modify what you want them to. See also **appositive; dangling modifier; only; this**.

MKS units. See **SI units**.

mole. The unit of quantity of matter, formerly called a *gram-mole*. The *kilogram-mole* is obsolete.

molecular weight. See **atomic mass, atomic weight; mass, weight**.

mouse. When a mouse is a little thingamabob attached to a computer, what is its plural? I have heard both *mice* and *mouses*. If you are in doubt, I recommend that you emulate the man who did not know the plural of mongoose; write

> Please ship one mouse.
> P.S. Well, while you are at it, send two.

Mrs. Young's law. Science as we know it would not exist if it weren't for masking tape.

multiple. Except in certain stock phrases like *multiple reflection interference*, this should almost always be replaced by the simpler and less pompous *many* or *several*, depending on meaning. See also **as many as; numerous**.

multiplication. See **symbols**.

myself, yourself. Not synonyms for *me, you.* It is not correct to write

> The research was performed by White, Stewart, and myself.

The correct form is

> The research was performed by White, Stewart, and me.

See **first person**.

Another collector's item:

> Sounds that myself as a classically trained musician is used to hearing.

N

naked decimal point. A decimal point is a little thing; it can easily get lost. Do not leave your decimal points *naked* but precede them with a digit. That is, when reporting numbers less than 1 write them with a leading 0: 0.7071, not .7071. See also **numbers**.

namely. Not as bad as the lawyer's *to wit* but almost always superfluous. In

> The cancer cells lose one of their major characteristics, namely, the ability to divide continuously,

delete *namely*:

> The cancer cells lose one of their major characteristics, the ability to divide continuously.

names. Don't butcher them. Kirchhoff, for example, has two *h*'s; Planck is spelled with a *c*. The most troublesome are names that end in *s*, like Pockels or Huyghens. The cell developed by Pockels is not a *Pockel's cell*; it is a *Pockels cell*. The *Huyghens construction* (or *Huyghens's construction*) is not *Huyghen's construction*. If you have any doubt about a name, look it up. Neither a word processor nor a copy editor is likely to help with misspelled names.

In this regard, I prefer to spell words like *Gaussian* or *Lorentzian* with capital letters, in recognition of the people for whom such words were coined.

near. Why *near-perfect, near-coherent*, and *near-ambient temperature* instead of *nearly perfect, nearly coherent*, and *nearly ambient temperature*? See **adverb; -ly**.

neodymium. Not neodyNium, though you often hear it pronounced that way.

neologism. Inventing new words or, sometimes, giving old words new meanings. You may share your computer with someone, but that is different from *time-sharing* it. Time-sharing is a neologism that implies you are using the same computer at the same time; that is different from ordinary sharing.

A colleague of mine says that we need neologisms to replace the words we are destroying with **euphemisms**. Besides, if no one uses neologisms, where will our new words come from? Use neologisms when they are clear, but avoid making up your own unless you have a sound reason and know you will be understood. See also **retronym**.

new innovation. All innovations are new.

nice. Has almost no value except for damning with faint praise. *Nice* is best used to mean *precise, exact,* as in *a nice fit, a nice problem, a nice distinction. Nice* should be avoided when used to mean, well, *nice.*

no way. A silly expression that has no meaning and adds no strength to your writing. It often leads to cumbersome **double negatives** like

> I was the first woman in America to rise to the editorship of a major newspaper, and there was no way I was not going to bring other women with me.

This sentence should have been rewritten something like

> As the first woman in America to rise to the editorship of a major newspaper, I was determined to bring other women with me.

nominal. Really means *in name,* not *approximate.* For example, a two-by-four is *nominally* two inches by four inches in cross section, but it is in reality *approximately* 1.5 inches by 3.5 inches (and shrinking). The distinction between *nominal* and *approximate* is worth preserving.

When you use nominal, make sure it modifies the right word.

> The plate is 210 mm in nominal diameter

is wrong: it is not the diameter that is nominal but the 210 millimeters. The sentence should be rewritten

> The plate is nominally 210 mm is diameter.

Use *nominally* when an adverb is required: *nominally 50 micrometer core,* not *nominal 50 micrometer core.*

non-. This is a **prefix**, not a word, and cannot stand alone. It must be attached to a word, usually without a **hyphen**: *nontechnical, nonionic, nonstandard.* In addition, there is no need to use a hyphen to prevent doubling a consonant: *nonnegative, nonneutral.*

When and how to use a prefix like *non-* can be tricky, especially when jargon is involved. Recently, a friend was working on a paper that used the psychological term, *status assigner.* She asked how to hyphenate its opposite, *non-status assigner*, someone who does not assign you status. Grammatically, the term should be *status nonassigner*, or perhaps *not a status assigner*, but that is not the jargon of the psychologist. After three days, I opted in favor of the hyphen, thinking that it would remove the prefix to arm's length and imply that it negated the entire phrase rather than just *status.*

The electrical engineers are one up on the psychologists, however. The term *dispersion unshifted*, for a fiber that has not had its dispersion properties shifted or altered, has gained acceptance. Although it is clumsy, it is lots better than *undispersion shifted.*

Be careful not to use *non-* to signify the opposite of a word that already has a perfectly good opposite. *Nonnegative* is all right, because it is not the same as *positive*; the nonnegative numbers include zero. But a *non-high-*T_c *superconductor* is just a *low-*T_c *superconductor.*

Non- has also become something of a vogue word. *Nontrivial* as an ironic synonym for extremely difficult may be all right, but do not invent *non-enriched* or *non-successful* when the synonyms *unenriched* and *unsuccessful* already exist. See also **vague words, vogue words**.

nor. Don't forget to use *nor* with *neither* or other negative sentences. For example,

> Neither Randi, Maddox, or Stewart has a background in immunology

would have been preferable as

> Neither Randi, Maddox, nor Stewart has a background in immunology.

Similarly, consider *nor* in a sentence like

They did not believe it was an asteroid nor any other extraterrestrial factor.

normalcy. Normality.

not without. Usually used when you want to be ironic and therefore usually out of place in a technical paper.

The method is not without its problems

really means that the method is plagued with problems. See **double negative**.

note that. Like *it is to be noted that*, these are usually wasted or **dead words**. It is, however, appropriate to use *note* or *note that* when you really have something you want the reader not to overlook. For example,

Note the difference of horizontal scale between the two figures

is specifically intended to call attention to a fact that the reader might not notice and is a lot stronger than

There is a difference of horizontal scale between the two figures.

But with this kind of exception, avoid sentences like

Note that the ambient temperature was 20°C

and simply state the fact:

The ambient temperature was 20°C.

See also **exclamation point; it**.

noun as adjective. See **adjective; inanimate possessive**.

noun extender. A meaningless adjective attached to a noun like filler to a hamburger. *Game plan*, for example, means no more than *plan*, nor does *action plan*. These expressions add nothing but syllables to your writing. See also **redundant expressions**.

now. An adverb, not an adjective. Avoid phrases like *the now-available report*. Instead, if the timing is important, use *the report*, *which is now available*. Otherwise, just say *the report*. See also **above, below; then**.

nth. Written *n*th. A perfectly good word, and the only one I know that has no vowels. You do not need to superscript the *th*, and you do not need to write *n-th*. Ditto with *i*th, *j*th, or *m*th, which, I suppose, are also words, but do not show up in the dictionary.

nuclear. Nu-clee-ar, not nu-cu-ler. A *nuclear event* is a euphemism for a hydrogen bomb explosion.

number of. See **singular or plural**.

numbers. Many countries use a comma, not a dot, as the decimal marker. In numbers with lots of digits, therefore, separate the digits into groups of three and leave spaces, not commas, between groups. Count both ways from the decimal point and write, for example, 299 792 458 or 5.670 51, but not 299,792,458 or 5.670,51. Leave out the space when only four digits are involved (1066, not 1 066), unless the number appears in a table with longer numbers and the space is needed for appearance.

Not all publications use this style, which reflects modern metric practice and is consistent with the International System or **SI units**. As a courtesy to the editor of such a publication, follow its **style manual**. But in your own reports or in camera-ready material, please stick with the International System. See also **naked decimal point**.

numbers in sentence. There is an arbitrary rule that you spell numbers less than or equal to ten; some people add to this round numbers such as twenty, thirty, and one hundred. (Others spell **unity** and **zero** as well.) In technical writing, the rule applies only to numbers used to indicate quantity. Thus, we write *two, 40 kHz transmitters*, but *a factor of 3*. Seeing *a factor of three* in a manuscript always makes me want to ask, A factor of three what? Similarly, I would not write *Frame one* any more than I would write *Figure one*; write *Frame 1*. *Day one* is silly and deserves no further comment.

Some publications that require hyphenation would hyphenate something like *30-cm radius*, since *30-cm* is a compound modifier.

It is not customary, however, to hyphenate a construction like *30 cm long*, even though *30 cm* is an adverbial phrase that modifies the adjective *long*.

I would not, however, write *85 500 kiloton weapons*. Since, in modern metric usage, the delimiter for grouping the digits in large **numbers** is a space (not a **comma**), this could be understood as either *85 weapons of the 500 kiloton variety* or *85 thousand five hundred weapons of the kiloton variety*. It would be best to rephrase as *85 weapons that deliver an energy of 500 kilotons*. Alternatively, a comma would also relieve the ambiguity, provided that there was a space after it: *85, 500 kiloton weapons*. A **hyphen** will not work here, incidentally, if you use the space as the delimiter: *85 500-kiloton weapons* could still be read either way.

When you express a range with a **hyphen**, do not add unnecessary spaces. For example, write *Fig. 6(a–d)*, not *Fig. 6(a – d)* or *Fig. 6 (a – d)*; *150–200 years*, not *150 – 200 years*; and *140 kHz–6 MHz*, not *140 kHz – 6 MHz*. (The hyphen, like most other punctuation marks, rarely, if ever, requires a space on both sides.) See also **between; equations; from; naked decimal point**.

numerous. Like **multiple**, a pompous synonym for *many*. Use *numerous* only when you mean *a very great many*. Do not write

> We repeated the experiment numerous times

when you mean three, four, or even a couple of dozen. Write instead

> We repeated the experiment several times

or

> We repeated the experiment many times,

depending on your mood. Reserve *numerous* for sentences like

> The fact of evolution has been demonstrated numerous times,

where the number is so large that it is practically uncountable.

O

obeys an equation. An interesting bit of jargon. We always talk as if the electron knows the equations of electromagnetism and therefore follows the appropriate path. In reality, the equation may or may not describe the path of the electron. Never let a philosopher catch you using this expression.

obvious. See **trivial**.

observed to, seen to. Usually **dead words**, as in

> The water temperature was observed to vary by 0.7 K during a run.

It would be better to use the **active voice** and write

> We observed the water temperature to vary by 0.7 K during a run,

but, since the observer is less important in this sentence than the observation, it would probably be best to write

> The water temperature varied by 0.7 K during a run.

of course. This little phrase implies that everyone knows what you are about to say. If everyone knows it, you may not have to remind us. If everyone does not know it, you may be showing off by using *of course*. For example, in a popular science magazine, I saw the parenthetical sentence

> (Factoring, of course, is finding numbers, or factors, by which another number can be divided evenly.)

In the context, it was proper to define factoring, but *of course* was gratuitous and implied *I know it, even though you don't*. The sentence would have been less supercilious written

> (Factoring is finding numbers, or factors, by which another number can be divided evenly.)

Avoid *of course* unless you are really sure the information is a reminder.

of mine. *A colleague of mine* is an interesting redundancy. You could argue that it should be *a colleague of me*, but it isn't. We say *a colleague of mine* or *a subordinate of Shirley's*.

on, onto. See **in, into**.

one. See **first person; numbers in sentence; unity.** The British use *one* as a personal pronoun, but Americans find it stilted at best. Avoid it.

one and the same. The same. See also **redundant expressions**.

one of us. In a paper with several authors, you will often see

> One of us (C.D.B.) has shown earlier that the form of the curve is quadratic.[5]

This always seems pompous to me. Just write

> Brown has shown earlier that the form of the equation is quadratic.[5]

Similarly, in **acknowledgements**, do not write

> One of us (M.Y.) is grateful to the American Optical Company for its support.

Instead, just write

> M. Y. is grateful to the American Optical Company for its support.

one single. *One* or *a single.*

ones. In the sentence,

> The data for the anodized connectors are similar to the data for the unanodized ones,

the author was evidently afraid to repeat *connector* and wrote something a lot worse than what he was trying to avoid. He could instead have repeated *connectors*,

> The data for the anodized connectors are similar to the data for the unanodized connectors,

or left it out entirely,

The data for the anodized connectors are similar to the data for the unanodized.

See also **false elegance**.

only. This word is probably misplaced more than any other modifier.

We only investigate this simple case here

means we do nothing but investigate; we draw no conclusions. This is so because the modifier *only* precedes the verb *investigate*. It should precede *this simple case*. We investigate—what? Only this simple case. The following sentences all have different meanings because of the placement of the adverb *only*.

Only we investigate this simple case here.
We *only investigate* this simple case here.
We investigate *only this simple case* here.
We investigate this simple case *here only*.

See also **misplaced modifier**.

only just. Only.

op. cit. The abbreviation for the Latin phrase, *opere citato*, in the work cited. See **ibid**.

operative. I have never been able to figure out what *operative* means, except in phrases like *CIA operative* or *The Continental Op* (a book by Dashiell Hammett). But what does it mean in

The operative word here is greed?

A better phrase might be *the key word* or *the important factor*. *Operative procedure* is a clumsy phrase for *operation*.

or. I sometimes wonder what verb to use with a compound subject that uses *or*.

Either the flashlamps or the rod *was or were* damaged?

The rule is to make the verb agree with the second (or the last) noun or pronoun in the subject:

Either the flashlamps or the rod was damaged,

but

Either the rod or the flashlamps were damaged.

The trouble is that the rule doesn't always work. The first example here doesn't sound so wonderful, but what about

You or I am going to do it?

Dreadful. Use the rule when it works; use your ear when it does not. Write

You or I are going to do it

if that sounds best to you.

organization. Most full-length technical papers are organized according to the schemes I discussed in Part I and under the entry **outline**. Very short papers, some reports, and articles in trade or popular magazines may be organized differently. In addition, you may sometimes have to decide how to organize sections within the body of your paper.

There are several common organization schemes: *order of importance, order of descending importance, chronological* (or, sometimes, *geographical*) *order*, and a combination of orders.

A paper that follows the ANSI recommendation, Introduction, Methods, Materials, Results, and Discussion, is apt to be written in order of importance, also called *order of increasing importance* or *climactic order*. Order of importance is often appropriate for a very short paper or a memo, for example. In these, the conclusions or recommendations are left for the very end of the paper, and the paper follows a didactic or deductive organization, leading us logically to the conclusion.

Longer papers, I think, should not be written in order of importance. Rather, the introduction should contain at least an indication of the conclusions or recommendations of the paper. See **getting started**. The remainder of the paper might then be written in order of importance and lead us to the final conclusion. It is perfectly acceptable, however, to state your conclusion very early in the paper, perhaps in the opening paragraph. A technical paper is not a detective story, and you do not give anything away by telling us where you intend to end up.

Popular or trade magazine articles can be very effectively written in order of importance, particularly if the story has a real climax. But it can be overdone: Recently, I gritted my teeth through an entire article about a man who claimed to have invented a perpetual motion machine. Not until the very end of the article did the author make his essential point, that a simple calculation could easily debunk the "invention." The reader who did not make it to the end may well have come away with the impression that there was something to the "inventor's" claims.

Order of descending importance is often called the *inverted pyramid* style. This style should generally be avoided in technical papers; it is really designed for daily newspapers, where the editor who wants to cut a few paragraphs may not have time to rewrite the article. Within sections, however, your paper may well be organized in order of descending importance. For example, in a section on materials or apparatus, you might describe the most important facts first and leave incidental facts to the end of the section. The whole paper, however, should not be organized according to order of descending importance.

Chronological order is sometimes appropriate for a review paper or a report. In addition, a mathematical or theoretical paper may be organized in what is essentially chronological order; that is, a proof or exposition is ordered sequentially, not necessarily the way you discovered it, but perhaps the way you would present it at a lecture. Likewise, instruction manuals or descriptions of a process or experiment may well be organized in the order in which certain operations are carried out. Sections of a paper, such as those that deal with the methods, the experiment, or the background, may well be written in chronological order, even though the paper as a whole may not be organized chronologically. Again, chronological order may be most suitable for trade or popular magazine articles, where the history or the participants are at least as interesting as the outcome.

If order of descending importance is the inverted pyramid, then we could, following Blicq, call order of importance the *pyramid* style. The most effective papers, it seems to me, do not follow the pyramid style; rather, they follow a combination of orders. They have a strong, effective introduction and then, possibly, follow order of importance. Using the pyramid analogy, we could visualize this as a pyramid with a block or an inverted pyramid on top of it, or, as Tichy says, a carafe or an hourglass.

Many full-length papers are organized in the carafe style. They are written so that a nonspecialist can get a lot out of them by reading only the introduction and the conclusions. In some ways, this is an ideal to strive for. See also **abstract; conclusions; introduction**.

Orwell's laws. See **barbarism; dead words**.

orientate. This is probably what linguists call a back-formation from *orientation*. The correct work is *orient*.

out there. Where is *out there*? This silly phrase barely qualities as jargon. Don't write

> There are billions of galaxies out there.

Be specific; write

> There are billions of galaxies in the universe.

See also **vague words, vogue words**.

outage. A barbarism for *power failure*.

outline. I have never been able to work from an outline, and I understand that many good technical writers do not do so either. (In high school, I used to leave a space for the required outline and fill it in after I had finished my essay. Now they know.) I find it much better to review my material until I know and understand it thoroughly, and then sit down and tell the story. I find the beginning the hardest thing to write; once I have a beginning, the rest is easy. See **getting started**. Sometimes, unfortunately, the beginning takes a lot of **procrastination**. After I am done, I go back and fill in the **abstract**; possibly this is a vestige of my high school English career.

Still, most technical papers are organized along similar lines: **title, abstract, introduction**, theory, experiment or apparatus, observations, summary, recommendations or **conclusions**, possibly an **appendix**, and **references**. Some sections may be left out, and others may be subdivided, or the introduction may include a summary. The ANSI recommendation—introduction, materials, methods, results and discussion—combines several of these sections. These outlines or something like them are good ones to work from, even if they are only in your head. See also **organization.**

output. See **input, output**.

over against. A redundant way of saying *against*:

> There is no longer an established norm over against which we are measured

should have had *over* left out, though it would have been more direct as

> We are no longer measured against an established norm.

over and above. *Over* or *beyond.*

own personal. Own.

P

page charges. Most journals published by scientific and engineering societies assess page charges for publishing in their journals. These page charges offset the cost of publishing the journals and help make the journals affordable to the members. (*Proprietary journals* are those published by commercial publishers for profit. Individuals rarely subscribe to proprietary journals, and they cost your library several times more per page or per 1000 words than society journals.) I consider the cost of publishing an article part of the cost of doing research and always urge people to pay the page charges, even though they are to some extent voluntary.

paragraph. One thought per sentence, and not too many per paragraph! Each time you feel yourself changing gears or introducing another topic, begin a new paragraph. Better to have too many paragraphs than too few. Paragraphs break up the monotony of a gray page and give the eye some reference points, so that readers know where they are on the page.

Try to make all your paragraphs begin with a *topic sentence* or some other indication of the subject. This will help the reader who is skimming for the more important material.

You can emphasize a point by writing it into a one-sentence paragraph.

parentheses. Parentheses are a useful way of inserting nonrestrictive words and phrases (or explanatory phrases). (Parentheses can also be used to call attention to a nonrestrictive phrase.) Punctuating parenthetical remarks is sometimes tricky. When an entire sentence is inserted between parentheses, the period goes inside the parentheses; otherwise, the period belongs outside the parentheses. Some writers mistakenly use commas and parentheses together, (like this). For example,

> This is in contrast to heterodyning, (frequency translation) when the signal contains phase noise.

In this sentence, *frequency translation*, though parenthetical, is in apposition to *heterodyning* and should be set off with either commas or parentheses, but not both. This author needs to decide whether to use parentheses,

> This is in contrast to heterodyning (frequency translation) when the signal contains noise,

or commas,

> This is in contrast to heterodyning, or frequency translation, when the signal contains noise.

Parentheses can also be used to drive the reader to distraction.

> The former (latter) is the distance at which amplitude (intensity) falls to $1/e$ of its value on the axis.

This atrocious construction seems to be spreading like a weed and is badly in need of some herbicide. It should be revised to something like

> The former is the distance at which the amplitude falls to $1/e$ of its value on the axis, and the latter is the distance at which the intensity falls to $1/e$.

It would be best, though, not to use *former* and *latter* at all, and instead write

> L_1 is the distance at which the amplitude falls to $1/e$ of its value on the axis, and L_2 is the distance at which the intensity falls to $1/e$.

Never mind that you may have used L_1 and L_2 in the preceding sentence; your goal is clarity, and clarity is best served by repeating the symbols. See also **false elegance; last, latter**.

A related form that I find very difficult to read in a single pass is something like

> The plus (minus) sign is used for distances to the right (left) of the lens.

It is left to my ingenuity to figure out that the quantities in parentheses may be read in place of the words that precede them (respectively?). Try reading this sentence aloud and seeing if you can understand it. You cannot. The sentence should be rewritten

> The plus sign is used for distances to the right of the lens and the minus sign for distances to the left.

A construction that is not much better is

> A scoop(s) removes the balls from the grooves,

which should be rewritten

> One or more scoops remove the balls from the grooves

to avoid the ambiguity introduced by the parenthetical *s*. See also **or**.

Some writers like to use parentheses to offset the numbers in a list: 1) or (1), for example. When the numbers always appear at the beginning of a line, the form 1) may be acceptable in place of 1., but when the numbers are inserted as part of the sentence, it is clearer to enclose each number with two parentheses:

> The parts of a technical article are (1) Introduction, (2) Materials, (3) Methods, (4) Results, and (5) Discussion.

Parentheses, like lions and quotation marks, run in pairs.

The use of parentheses in mathematics goes without saying, but one form I have seen causes confusion: expressing the limit of error in the form 2.997 792 458(4). Although this form may become standard, it is not obviously equivalent to 2.997 792 458 ± 0.000 000 004. This form is especially odd when a decimal point appears near the last digit, as in 105.4(16), which means 105.4 ± 1.6. If you want to use this form, define it the first time you use it in a paper.

partake, participate. *Partake* usually means *eat*, but it can be used in the sense of *partaking of a friend's hospitality*. It is not a good synonym for *participate*. Usually you participate *in* something, such as a game, but you partake *of* something.

partially. Partly.

participle. When used as a modifier, the *-ing* form of a verb is known as the *present participle*; the *-ed* form is the *past participle*. The present participle is vastly overused in technical writing; it should usually be replaced with a *that* or *which* clause. See **dangling modifier; misplaced modifier**.

It is incorrect to use a present participle as a modifier when the main verb of a sentence is in the past **tense**.

Pascal's law. If I had more time, I could write a shorter letter. Amos Kenan put it a little differently: I haven't got enough time not to make mistakes. In short, don't agonize over the last data point; write your paper. Similarly, don't agonize over the paper; submit it.

passive voice. Technical papers can be written in the passive voice. This is done by some writers with grace and skill. Possibly, technical papers are written in this way to emphasize that their results are universal and could have been obtained by any scientist. It is felt that the author should therefore be separated from the work by using the device of the passive voice.

Now consider the preceding paragraph written in the **active voice**:

Some technical writers use the passive voice with grace and skill. They probably write that way to emphasize that their results are universal and that any scientist could have obtained them. They believe that the author should separate himself from the work, and they use the device of the passive voice to do so.

Fifty-four words, as opposed to 63 in the first version. You can almost always save space by writing in the active voice, whether you use the **first person** or the third person.

> In a recent issue, the use of lasers in medicine was discussed,

for example, shortens nicely to

> A recent issue discussed the use of lasers in medicine.

Particularly bad examples are *Use was made of* and *used by us*, which shorten to *We used*. For instance,

> The psf of the system used by us is reduced by a factor of 1.4

would have been clearer written

> The psf of the system we used is reduced by a factor of 1.4

or

> The psf of our system is reduced by a factor of 1.4.

Sometimes the passive voice changes the emphasis of the sentence or leaves the sentence with no emphasis.

> This argument, too, is disposed of by Churchland

is bland and emphasizes Churchland, if anything. What we are interested in, however, is her result:

> Churchland disposes of this argument too.

In an instruction manual, I came across this interesting bit of advice:

> Do not use until instructions have been read.

Well, they have been read; they just haven't been read by me.

The passive voice often prevents you from saying what you really mean. It is most appropriate when the actor is unknown or irrelevant, say, when you are describing some apparatus or making a statement about some universal fact. For example, you might write

> Light of these wavelengths may be implicated in senile macular degeneration

if you are talking about the overall effect of sunlight on the eyes. But there is a difference between

> Near infrared radiation is transmitted efficiently

and

> Many sunglasses transmit infrared radiation efficiently,

because the second example emphasizes that the sunglasses are the culprit.

Another unfortunate side effect of always writing in the passive voice is reverse word order:

> Affected would be the northern parts of the U.S. and Canada.

This construction removes the emphasis from Canada and the

northern parts of the U.S. (which is, by the way, unambiguous) and leaves it—nowhere. The sentence should be

> Canada and the northern parts of the U.S. would be affected

or, if you want to strengthen your conclusion,

> Canada and the northern parts of the U.S. will be affected.

Unless you have a very good reason, avoid the reverse word order found in the common construction

> Also plotted in Fig. 6 is the unperturbed wavefunction.

This sentence, like many others, would sound better if it began with its subject, whether the subject is wavefunction:

> The unperturbed wavefunction is also plotted in Fig. 6,

or Figure 6:

> Figure 6 also shows the unperturbed wavefunction.

The last example is written in the active voice and has the most natural word order: subject, verb, object. It would be unduly rigid to prescribe beginning every sentence with its subject, but generally adhering to such a rule would improve a great deal of awkward writing. See also **it; dangling modifier; misplaced modifier; there; word order**.

percent. I know it is an abbreviation for the Latin phrase *per centum*, but most people spell it *percent*, one word, no period. Some publications use the percent symbol %, whereas others prefer the complete spelling. My preference is to spell percent if I am spelling the names of the units, but to use the percent sign when I am abbreviating the units. In tables and figures, however, the percent symbol is generally acceptable. Always use digits, not words, before percent or %: 10 percent or 10%, but not ten percent.

When something goes up 10 percent, it goes up 10 percent, not 10 *percentage points*. This clumsy expression probably derives from the Dow Jones average, which is said to go up (or down) by *points*, even though the units of the Dow Jones average are really dollars.

If a *rate* changes from 10 percent to 11 percent, it goes up 1 percent or *by* 10 percent, depending on whether the incremental change of 1 percent or the relative change of 10 percent is important. There is no need to use the phrase *percentage point* to distinguish one from the other.

Incidentally, an increase of 500 percent is a factor of 6, not 5.

performance anxiety. Stage fright.

period. There is usually no need for the modifier *time* in *time period.*

permissivist, prescriptivist. A *prescriptivist* is a person who believes that language follows a fixed set of rules that never change. Doctrinaire prescriptivists may not quite believe that the original meaning of a word is the only acceptable meaning, but they will certainly resist other meanings until they are firmly entrenched. The prescriptivist is the fundamentalist of style.

The trouble is that language is not revealed truth; it is what people say and write. The prohibition against splitting infinitives, for example, derives from the time when European languages were treated almost as if they were corruptions of Latin. In Latin, the infinitive is one word; therefore (?), the infinitive in English must likewise be treated as one word. (Never mind that English is not a Romance language, nor that the English infinitive is two words, nor that native speakers commonly use split infinitives.) Similar arguments can be made about the prescriptivist's prohibition against ending a sentence with a preposition or the prescriptivist's distinctions between *that* and *which* or *shall* and *will.*

The temptation is to throw away all prescriptions and argue that anything people say goes; this is the approach of the *permissivist.* I argue to the contrary. As Mordecai M. Kaplan said in another context, the past has a vote, but not a veto. So what if *awful* once meant awe-inspiring? Today it means very bad; we use *awesome* when we mean awe-inspiring. Past usage is important, but you have to consider the recent past as well, and you have to be sure you are listening to educated speakers and writers before you decide what goes.

person. *The New York Times* called the effort to root out the suffix -man *No-men-clature.* Needlessly sexist terminology, whether real or imagined should be altered, but *grandparent clause*, for example, is historically inaccurate, and *birth attendant* is awfully clumsy for midwife. After *woman* is changed to *woperson*, how long will it be before someone notices that per*son* contains the masculine syllable *son*? See also **gender, sex; he or she; man, -man**.

phenomenon. *Phenomena* is the plural; you cannot write *a phenomena*. When you want the singular, use *phenomenon*. See also **agenda; criterion; data**.

plagiarism. A wag named Mizner once remarked

> If you steal from one author it's plagiarism; if you steal from many it's research.

S. J. Perelman said

> Mediocre writers borrow; the great ones steal,

and no less an authority than William Inge defined originality as undetected plagiarism.

Still, quoting someone's words precisely without attributing the words to the original author or using quotation marks is plagiarism. So is paraphrasing someone's idea without attribution. Publishing a plagiarized document or even a plagiarized fragment may be illegal if the **copyright** has not expired. It is certainly unethical if it is done consciously.

Literary *allusions* are not common in technical writing. They are different from plagiarism, in that they are clear references to earlier words and are intended to be recognized. For example, Abraham Pais's wonderful biography of Einstein was called *Subtle Is the Lord*; this was an allusion to Einstein's own words, "Subtle is the Lord, but he is not malicious."

Inadvertent plagiarism is almost unavoidable: Sometimes you may read a sentence or a phrase, internalize it, think it is your own, and eventually repeat it in print. Such plagiarism, limited to a few words or phrases, is insignificant. See also **references**.

player. When Shakespeare wrote

> All the world's a stage, and all the men and women merely players,

I don't think he meant his words to be taken literally. Why, then, do some authors write

> User-friendly imaging is on the horizon, thanks to such players as Apple Computer and Sun Microsystems?

Important people, corporations, and nations are not players; they are important people, corporations, and nations. *Players* should always be replaced with a more specific term (unless you are a sports writer and you *mean* players). In this example, write

> User-friendly imaging is on the horizon, thanks to such corporations as Apple Computer and Sun Microsystems.

See also **vague words, vogue words**.

plethora. An excess, not a lot, and a rather literary word that you would not expect to find in scientific papers. Yet I have recently seen it used in a couple of places where I suspected that the author really meant a lot; for example,

> The remarkable discovery of Bednorz and Müller produced a plethora of experimental studies.

Moral? If you don't know what a word means, look it up. See **false elegance**.

plurals. When there is a choice, I prefer the English plural; then you do not have to know that *platypuses* and *octopuses* are correct because the endings are Greek and not Latin, nor that the plural of rhinoceros is *rhinoceroi*. Similarly, it is startling to read

> Vartanyan has been the chief apologist for Soviet psychiatry in international fora,

when most writers would have used *forums*. Finally, I prefer *indexes* and *appendixes*, especially when an appendix is a part of a book.

What about compound nouns like *mother-in-law*? The books say that the plural is *mothers-in-law*, but the man in the street says *mother-in-laws*. If you think he's wrong, then what is the possessive?

On a scientific note, the dictionary says that the plural of *apparatus* is either *apparatus* or *apparatuses*. *Apparatuses* is kind of clumsy, but it sure beats the pompous Latin plural, *apparati*. (*Surely* would be correct here, but *sure* is idiomatic.) The plural of *infinite series* is *infinite series*; you can't write *serieses*, though I have heard this form in speech. Most other words whose singulars and plurals are the same are names of animals and less frequently of plants, and may be listed in a dictionary.

plus. *Plus* indicates addition; it is not a synonym for *and* or *in addition*. I'll accept *a big plus* but not *Plus, it has a color monitor*. See also **and (comma), but (comma)**.

point in time. Usually part of a silly expression like *at this point in time*, which means *now*. See also **time frame**.

possessive. See **apostrophe**.

postal code. The Post Office's abbreviations are for making it easier to sort mail; reserve them for addressing envelopes. I live in Boulder, Colorado, not Boulder, CO.

practically. Besides meaning *in a practical manner*, *practically* means *for practical purposes*. Colloquially it means *nearly*, but I would not use it in this sense in writing. *Practically 100 micrograms*, in a formal paper, would better be written *nearly 100 micrograms*.

preceding. One *e*, unlike *proceeding*. Often can be replaced by the simpler *last*.

precision and accuracy. Many metrologists make a distinction between *precision* and *accuracy*. If we think in terms of a ruler, *precision* is a measure of how finely divided the scale is—centimeters or millimeters, for example. *Accuracy*, on the other hand, is a measure of whether the ruler is the right length. A ruler that has a significant error in its length will be inaccurate no matter how precise or finely divided the scale.

In a real measurement system, *accuracy* usually describes *systematic errors*, including instrumental errors, whereas *precision* usually describes *random errors*, though sometimes these associations may be blurred.

precision measurement. Precise measurement. A *precision measurement* would be a measurement of precision.

predicate adjective. See **linking verb**.

predominant. Not predominATE. *Predominate* is a verb.

prefix, suffix. These are not words; they are particles like *un-*, *non-*, *-ly*, or *-ful*. Prefixes, in particular, cannot stand alone; they must be attached to words: *nonnegative*, not *non negative*. The hyphen is usually considered unnecessary unless the addition of the prefix would create a doubled vowel, *electro-optic*, or an unusual or ambiguous construction, *un-ionized* or *co-worker*. I would also hyphenate a word that used two prefixes sequentially: *Acousto-electro-optic effect*, not *acoustoelectro-optic effect* nor *acousto-electrooptic effect*. In addition, a hyphen is helpful before a capital letter, *non-Euclidean*, or between an *i* and an *e*, *anti-evolutionist*. More under **hyphen**.

Suffixes almost never require hyphens, unless you are using a made-up word, such as *comma-kaze* or *term-ite*, or doubling a vowel.

preliminary experiment. An experiment that we haven't done very well, are not terribly confident about, but have no intention of repeating.

preliminary to, prior to. Before.

preposition. A preposition is something you can very well end a sentence with. Every native speaker does. See **permissivist, prescriptivist; split infinitive**.

present day. Today.

> The task is formidable, even on computers of the present day

or

> The task is formidable, even on present-day computers

is simpler as

> The task is formidable, even on today's computers.

Indeed, the adjective *present-day* can usually be replaced by *current, existing*, or *contemporary*. See also **the present**.

presently. Used to mean *soon*, but slowly slipping to *now*. What's the matter with the shorter words, anyway?

presume. Not a synonym for *assume*. To presume that something is true is to accept it as fact, usually because there is evidence, or at least lack of evidence to the contrary. When Stanley said, "Dr. Livingston, I presume?" he knew perfectly well whom he was talking to.

To assume something, on the other hand, means to accept it provisionally, for example, for the sake of argument. It is not correct to write

> We presume the sample to be random

when, in fact, the sample may well not be random. What you mean is

> We assume the sample to be random

for the purpose of this argument, even though we have not tested it and know that it really might not be.

principal, principle. *Principal* is a noun that means chief or the head of a school, or an adjective that means first in importance. It is related to the nouns *prince* and *principality*, which give a clue to its spelling. A *principle* is an axiom or a law and should not be confused with a principal (even though some principals I have known thought they were the law).

Do not spell *in principle* as *in principal*.

principle of parallelism. There really is a principle of parallelism; otherwise it would not be funny to say

> The troops were greeted by General Eisenhower and general confusion.

OK, so it's not very funny, but it makes the point that the ear wants to hear (in this example) a name after General Eisenhower, and it is startled when it hears *general confusion*.

In writing, the principle of parallelism dictates (sometimes too strongly) that you use parallel constructions under certain circumstances. The most common of these is a **list**, and all the items in the list should generally have the same form—all nouns, all participles, all sentences, for example. Thus, it is not considered good form to write

> Previously they had been institutionalized in jails, mental hospitals, or for drug detoxification.

Since *jails* and *hospitals* are nouns, the third item in the list should be a noun like *drug detoxification centers*:

> Previously they had been institutionalized in jails, mental hospitals, or drug detoxification centers.

Otherwise, the sentence may be rewritten

> Previously they had been institutionalized in jails or mental hospitals, or for drug detoxification.

In this version, the list has two items, both prepositional phrases.

Similarly, don't write

> We need the computer, the controller, the interface, and to get an expert to run them,

where the first three items are nouns and the last is an infinitive. Instead, write

> We need the computer, the controller, the interface, and an expert to run them

or, if you want to emphasize the need for an operator,

> We need the computer, the controller, and the interface, and we need to get someone to run them.

The sentence

> One technique of minimizing birefringence is to twist the fiber

would be better rewritten

> One technique of minimizing birefringence is twisting the fiber,

since *minimizing* and *twisting* are both gerunds.

Sometimes, however, you can carry the principle of parallelism too far:

> We have studied the data, as well as carefully checked the hardware.

Few would say *checked*; the word could arguably be *checking*:

> We have studied the data, as well as carefully checking the hardware.

If it sounds stilted, change it. (You could, however, argue that *as well as* should have been changed to *and*:

> We have studied the data and carefully checked the hardware.

Then *checked* would stand. The authors apparently wanted to emphasize checking the hardware and opted for *as well as* instead.)

In another example, the principle of parallelism is impossible to apply:

> Besides interacting with the beam, the nitrogen also affects weldability.

You cannot write instead

> Besides interacts with the beam, the nitrogen also affects weldability.

Why, therefore, would anyone insist on

> The nitrogen affects weldability, as well as interacts with the beam?

prioritize. An ugly word that, for some reason, smacks of **gobbledygook.** See also **-ize, -ization**.

private communication. Such references have no place in publications. They are almost completely useless unless I happen to know the writers and can call them freely. If you want to acknowledge someone for help or for an unpublished work, use the **acknowledgements** section. That's what it's for. If you insist on citing a *private communication*, at least include a complete mailing address, so that someone who does not have your connections can contact the person you have cited.

If you have access to an unpublished work—one that you are certain is actually going to be published—cite the authors, the title, and the journal in which you expect the article to appear Use **to be published** or *in press* only when the article has already been accepted; otherwise, use some phrase like *submitted* or *in preparation.*

If an unpublished article is not your own, be certain that you have the permission of the authors to use their results or cite their paper. At least one journal, *Science*, requires proof that you have such permission both for citations of unpublished work and for private communications.

problematic. Also problematical. Usually, *doubtful* will serve as well.

procedure. One *e*, even though *proceed* has two.

process. Often put in superfluously.

> The polishing process can be speeded up by adding a dilute etchant

is no better than the shorter

> Polishing can be speeded up by adding a dilute etchant.

Similarly,

> We are in the process of making the measurements

adds nothing to

> We are making the measurements now.

See also **vague words, vogue words**.

procrastination. I usually procrastinate for about a week before sitting down to write a paper and, when I can afford the luxury, for another week before preparing the final draft. It seems to be a very efficient way to collect my thoughts. Some of the processing that goes on during those weeks is apparently subliminal, but some seems quite liminal. When I cannot afford the time to procrastinate, the result is decidedly inferior or much harder to bring out. Far from being the thief of time, procrastination is a wonderful way to clean your lab and your office.

productization. For example,

> The productization of the MOCVD process has been the key.

Spare me! See **-ize, -ization; jargon**.

prophecy, prophesy. *Prophecy* is the noun. *Prophesy* is the verb. I once saw *prophesized* in a respectable mass-market science magazine (which subsequently went under).

proved, proven. The past participle of *prove* may be either *proved* or *proven*, but *proved* is becoming the preferred form:

> The theory of relativity was proved when Eddington measured the bending of light by a star.

The adjective is still *proven*, as in a *proven theory*. Why? See **Ciardi's law**.

provided that. The *American Heritage Dictionary*'s usage panel prefers *provided that* to *providing that*. I argue that it is not a **dangling modifier** when you write

> Provided that a tuned amplifier is used, the constraint is not apparent.

In this sentence, *provided that* is not quite the same as *if*. See **adverb** for a discussion of *sentence modifiers*.

pun. I once read somewhere that the pun is the lowest form of shredded wit. Still, I have put puns into technical papers, and I was not struck by lightening.

purpose. The putting in of extra words for the purpose of filling space is found to result in **dead words**. Indeed, *purpose* is very often a dead word. For example,

> For computational and color matching purposes, it is useful to know where the colors lie on the diagram

can easily be shortened to

> For computation and color matching, it is useful to know where the colors lie on the diagram.

Q

quantum leap. Since the changes in energy in quantum mechanics are very small, you might expect a *quantum leap* or *quantum jump* to be very small. *Quantum leap* has come instead to mean a sudden change, which occurs without passing through any intervening stages. The term is now used to mean any large, sudden change. It differs, I think, from a *sea change*, which is a rapid but continuous shift, as from east to west or, in politics, from right to left. Neither is a **redundant expression**. See, however, **noun extender**.

query. Not a bit different from the shorter and less formal *ask*, *query* should usually be replaced by *ask*, except perhaps in certain jargon phrases like *query a publisher*. Two of my three dictionaries, incidentally, give the pronunciation as queerie, not quehrie, and I consider queerie the preferred pronunciation.

question mark. This is, of course, used at the end of a question and, sometimes, in **parentheses**, to cast doubt on a word or phrase:

> He was a scientific(?) creationist.

The purist would argue that the question mark should be used only after a question or a quotation like

> Students always ask, "Why do we make that assumption?"

Nevertheless, I have seen question marks used effectively after implied questions like

> I have always wondered what was the origin of the uncertainty principle?

or, perhaps better,

> I have always wondered, What was the origin of the uncertainty principle?

Perhaps it is technically wrong to use the question mark in this way, without the quotation marks, but it is an effective way to draw attention to an implied question.

quotation mark. Words and phrases go inside quotation marks primarily when they are ascribed to someone other than the writer. Where and how to punctuate quotation marks is sometimes tricky.

> Lewis acknowledged in a telephone interview that "The real issue is how to come up with the money."

Here, the quotation should not have begun with a capital letter, because the conjunction *that* implies an indirect quotation or a partial quotation, and the quotation flows as part of the sentence. If I had wanted to show that I was using the whole quotation, I would have written

> Lewis acknowledged in a telephone interview, "The real issue is how to come up with the money,"

with a **comma** before the quotation and without *that*.

Some will argue that, in the original sentence, the period should have come after the quotation mark. In American typography, however, periods and commas are *always* placed inside quotation marks, irrespective of the logic of the construction. A **semicolon** (or, in another context, a **question mark** or an **exclamation point**) should, indeed, have been placed outside the quotation marks if it was not part of the quotation.

Certain usages, however, almost demand a comma outside the quotation marks, though the prescriptivist probably would omit it entirely:

> Given the question, "All oak trees have acorns. This tree has acorns. Is it an oak?", many adults insist it must be.

To me, the entire expression inside the quotation marks and the quotation marks themselves are in apposition to *question*, and were therefore properly set off with commas.

Some writers like to put single words into quotation marks, especially when they are using an odd word or an ordinary word slightly oddly. For example, once I called a single, spherical surface of glass a "len" (two of these form a lens). In that case, I decided to use quotation marks every time that word appeared, to avert possible confusion later.

But that was a made-up word. In a sentence like

> Published results are "weighted" according to a proprietary formula

or

> Poll takers rarely publish "raw" data,

weighted and *raw* are ordinary scientific words used properly; the quotation marks should have been dropped. On the other hand, in the same article I found

> The League of Women Voters set a threshold of 15% "in the polls" for participation in its televised presidential debates.

Here, the quotation marks are correct, because the thrust of the article was that "in the polls" is an ill-defined term.

In general, however, using quotation marks is like apologizing for your choice of words. Either find a better word or use your choice without embarrassment. If you insist on the quotation marks, use them only when you introduce the word. See also **right word**.

R

ragged right. Not **right justified**.

rather than. See **instead of, rather than; principle of parallelism**.

readability. See **Flesch's readability index**.

reason. Often a **dead word**, as in

> For reasons of simplicity, we did not distinguish between random and systematic errors.

Here, *reasons of* should be deleted:

> For simplicity, we did not distinguish between random and systematic errors.

See also **vague words, vogue words**.

reason why. The purist no doubt has a reason that this construction is improper, but I see no reason why you should not use it, even though it might be just a shade redundant.

reasonable. A much overused word. When I read

> In the absence of other information, it is reasonable to assume that the cause may have been economic,

I want to ask, Reasonable to whom? and wonder whether the author is trying to put one over on us. Likewise, what does *a reasonable size* or *reasonable agreement* mean? A prediction may be reasonable, but is it accurate? *Reasonable* should be reserved for people. See also **vague words, vogue words**.

recall that. See **dead words**.

recur. Not reOCcur.

redundant expressions. These are expressions that contain a word or words that add nothing to the meaning. For example, you would never say *a male man*; the word *male* is obviously superfluous, or redundant; all men are male; you are being redundant by repeating yourself. Here is a short list of common expressions that are less obviously redundant, but are redundant nevertheless.

Absolutely essential; action plan; close proximity (proximity means closeness); close vicinity; component parts; end result; exactly the same; fused together; future plans; game plan; general consensus of opinion (consensus means general opinion); line manager; mandatory requirement; new beginning or innovation (a new innovation would have to be your second innovation); only just; past experience; past history; performance objective; personal opinion; personal belongings; refer back; repeat or reiterate again (you repeat something again the third time you do it); road map (when used figuratively usually means just map); safe haven (a haven is a safe place); serious crisis (what crisis isn't?); subject matter; sum total; true (or actual) fact (if it isn't true, it isn't a fact); unexpected surprise. See also **noun extender**.

refereeing. Refereeing is a major part of publishing scientific papers. Without it, society presses would become little more than vanity presses to which authors pay the price of publication. Indeed, they have occasionally been accused of being just that. Therefore, refereeing a paper is a big responsibility and should be treated very seriously.

A number of years ago, H. J. Caulfield, then the editor of *Optical Engineering*, published a list of suggestions for referees. Here, with my glosses, is the gist of his remarks.

First, if you can review a paper, do it promptly. If you cannot, tell the editor immediately and, if possible, suggest another reviewer. Caulfield says, "Remember your annoyance with manuscripts that took months to review."

When you review the paper, do not feel compelled to find error or fault; many submissions display very high overall quality. When you review a poor paper, be as tactful as possible. Say what is good about the paper first, and make positive suggestions about what is bad and needs correction. Avoid implicit attacks on the writers rather than the manuscript. Try to distinguish between optional and mandatory changes. If you have to recommend outright rejection, be very specific about the reasons for your judgement. "Boring" and "old hat," Caulfield says, are not specific-enough reasons for an editor to reject a paper.

In addition, the referee should be certain that the paper reports something new and significant. (If the paper is a review article, it should present something not so new in a compelling or new fashion.) The article itself should make clear what is new, and the **introduction** should usually be clear to the nonspecialist. The referee should also, for example, make sure that the **abstract** is informative, not descriptive; that the references are complete and up to date; that the **figures** are clear and the **figure captions** complete; and that the summary or **conclusions** section is necessary.

I review three or four papers a year—the same order of magnitude I submit each year. Many of my comments have to do with style or clarity, and I write these right on the manuscript. In addition, I often supply a typed or written page or two of general or technical comments. Sometimes I ask a colleague to look over the paper as well, particularly if I am having difficulty or being very critical. Having once been the victim of (I thought) an unfairly critical review, I take the most pains with reviews that are generally negative, the better to avoid offending the authors. See also **manuscript preparation**.

reference. A noun, not a verb. You *refer to* something or *cite* it; you do not *reference* it. Similarly, do not write

> The signals are referenced to the frequencies of atomic oscillators,

but rather write

> The signals are referred [or compared] to the frequencies of atomic oscillators.

references. References should follow the **style manual** of the publication. Where possible, it is best to include the first and last page and the title of each reference. It is a great help to the reader to know both the subject and the length of a paper, and title and last page convey this information economically. Too often I have looked up a paper only to find that it was an abstract or a short communication, or a paper so narrow in scope that it was not relevant to me. Remember, the references are not there to show that you have done your homework (nor how erudite you are), but to aid the reader. Therefore, they should contain as much information as possible. See also **private communication; to be published**.

There are two main forms, the square bracket and the superscript, for citing a reference. The square bracket contains either a reference number or an author and date; it is usually inserted at the end of a sentence, before the period:

> . . . a recent study of the pinhole camera [7].

Or

> . . . a recent study of the pinhole camera [YOUNG, 1972].

A citation to several references looks better with only one pair of brackets:

> [7–9, 12],

not

> [7]–[9], [12]

or

> [7], [8], [9], [12].

The word *Reference* and the abbreviation *Ref.* are not usually used in text when you are citing a reference:

> See [7],

not

> See Ref. [7].

But see the style manual or check the journal to be sure.

The superscripted reference number should appear after the period or other punctuation:

> . . . a recent study of the pinhole camera.[7]

This is especially important when there are likely to be numbers or symbols in the sentence:

> a function of T.[2]

is a lot different from

a function of T^2.

I think it is also best, where possible, to put the reference at the end of the sentence:

... a recent study of the pinhole camera.[7]

rather than in the middle, where it breaks up the flow:

... a recent study[7] of the pinhole camera.

When it is necessary for clarity, as in a list of authors, the reference should, of course, be attached to the names. When you mention a reference in the text, use *Ref. 7* or *Reference 7*, according to the usage of the publication.

Make your references as specific as possible. Don't refer us to a whole book when your point is actually made on page 276 of that book. Instead, refer us to the appropriate pages or chapter of the book, unless your reference is intended to be very general.

If your reference list is numbered (as opposed to alphabetical), list your references in the order in which they appear in your text. If you need to cite a reference a second time, use the same number you used before; scientific or technical style permits a reference number to be repeated *in the text*. See, however, **ibid.** for a mention of scholarly reference style.

When you list your references, use the style of the journal to which you are submitting your paper. When I have my choice, I use the American Institute of Physics style:

3. M. Young, "Pinhole Optics," Appl. Opt. **10**, 2763–2767 (1972).

If you are allowed to cite titles, use a consistent **capitalization** scheme: capitalize only the first word of the title, or capitalize each word, irrespective of the way the original papers were capitalized.

When citing journal articles, give the author or authors, preferably the title, the name of the journal, the volume (and, if appropriate, the issue), first and last pages, and the year. When you are citing a trade magazine or a popular magazine that is not paginated consecutively from issue to issue, leave out the volume number and use the year and month instead:

> 7. Constance Holden, "Twins Reunited," *Science 80*, November 1980, pp. 54–58.

Put the title into **quotation marks** if that is the style of the publication.

When citing a book, give the author or authors, the title, the publisher and city, and year:

> 5. Ronald W. Clark, *Einstein, The Life and Times* (Harry N. Abrams, New York, 1984.)

Underline or italicize the title of the book if that is the style of the publication. For the precise format to use, see the publisher's style manual or "Information for Contributors" section, or check current issues.

Finally, do not confuse the authors with their article: not

> Equation (4) is derived in Morrison et al.,

but rather

> Equation (4) is derived in the paper by Morrison et al.,

or

> Equation (4) is derived by Morrison and his colleagues.

referred to as. Called. Reserve *referred to* for books.

regard, regards. You may give someone your *regards*, but you tell him *in regard to* something, not *in regards to*.

regime. This is beginning to be used in place of *region* and *regimen*, neither of which is a synonym. A *regime* is a political or social system (except in mathematics, where it is the range of values of a function). A *regimen* is a strict system of diet, exercise, and all that.

> The new accelerator will allow scientists to study this energy regime in detail

would be better as

> The new accelerator will allow scientists to study this energy range in detail.

reiterate, repeat. To *reiterate* is to *iterate again.* You cannot *re-iterate* something until the *third* time you do it. Similarly, to *repeat* something is to *say it again.* Usually, you don't repeat something again; you merely repeat it. To repeat something again is to do it three or more times. See also **redundant expressions**.

relative to. This is the preposition many use when they can't think of the right one. *I want to talk to you relative to your experiment.* What's the matter with *about*? Relative to this, avoid *relative to* at all costs. For example,

> Temperatures are incremented relative to absolute zero

meant

> The temperature is increased in steps beginning with absolute zero.

Similarly,

> Frank's comments relative to untapped markets are appropriate

should have been shortened to

> Frank's comments about untapped markets are appropriate.

See **vague words, vogue words**.

Relative to is correct when you are drawing a comparison or putting something into perspective, as in the sentence,

> These membranes have low permeability relative to commercial membranes.

This sentence makes clear that the membranes do not have low permeability on some absolute scale, but only with respect to the permeability of commercial membranes. It does not imply that the membranes have lower permeability than the commercial membranes. If that is what the authors meant, they should have written

> These membranes have lower permeability than commercial membranes.

See also **compared to; than**.

remotely piloted vehicle. Drone.

repeated measurements. Not *repeat measurements*.

repetition. See **false elegance**.

research. The regional pronunciation REEsearch (which accents the prefix and not the root) may be overtaking the more standard reSEARCH; possibly this is the influence of professional football, which has given us DEEfense and arthroscopic surgery. Still, I think that science journalists and researchers ought to use the more elegant pronunciation, reSEARCH.

respectively. Badly overused. In many cases, only an idiot would think that the entries were not in order. For example, in

> The horizontal and vertical components are denoted by H and V, respectively,

delete *respectively*.

On the other hand, a sentence like

> Mean loss coefficients are 0.35 and 0.21 dB/km at 1300 and 1550 nm, respectively

is hard to read. To figure out the loss at 1300 nm, I have to go back and read the sentence again. The sentence would be better and clearer rewritten

> The mean loss coefficient is 0.35 dB/km at 1300 nm and 0.21 dB/km at 1550 nm.

This form is much easier to understand without backtracking. See also **parentheses**.

result of the fact that. Because.

resumes. Without a good resume, you won't get the interview. Prepare your resume on a letter-quality printer or have a good typist prepare it for you. Your resume is the foot in the door, so do not spare any effort.

I prefer a journalistic style for resumes, though some people use sentences or paragraphs. One thing I suggest that is a little unusual is to start your resume with a headline, something like a title:

David Young
4289 Graham Court
Glen Cove, NY 11542
516-444-XXXX
MATHEMATICIAN SEEKING COMPUTER-RELATED POSITION

This headline takes the place of the *job objective* statement in many resumes. After the headline, you can put a couple of subheads if you like:

** Six years' experience in **computer sales and service**

** Author of **published computer program**

After the headlines, write a more-or-less conventional resume, with your education, experience, honors, and references detailed. List education and experience in reverse chronological order, with dates, titles, employers' names, and duties. For example:

EDUCATION. **University of Colorado**, majoring in math and French, 1985-now. Plan to graduate December 1989. Current **grade-point average, 3.4/4.0**.

EXPERIENCE. **User Friendly Computers**, part-time **salesman** and **instructor** of courses in Operating Systems, Word Processors, 1988-now.
Computer Works, part-time **salesman**, 1987–1988.
The Apple Polisher, a teachers' grade-book program published in 1987 by. . . .

HONORS. **Colorado State Science Fair**, second place in Engineering and three Special Awards for [name of project], 1985.

REFERENCES. Mr. Geoffrey Gordon, [address, telephone number]....

I do not recommend writing

REFERENCES available upon request.

Instead, find two or three people, preferably former employers or professors, who will agree to give you a good referral. Get their permission and put their names and telephone numbers on the resume. That way, employers who are interested in you will not have to waste time or create a delay by calling or writing you to get information you could have provided with your resume.

If you have any special skills, be sure to state them prominently on the resume, perhaps right after EXPERIENCE. For example:

ADDITIONAL SKILLS. Familiarity with Apple, IBM, and Atari computers; Microsoft Quick BASIC, assembly language, and FORTRAN; Word Perfect (a word-processing program) and D-Base (a spreadsheet program).

Avoid complete sentences like

I was responsible for organizing the undergraduate laboratories and supervised a faculty of 7.

Instead, write in a clipped, almost telegraphic style without the pronoun *I* or the auxiliary verb *to be*:

Responsible for organizing undergraduate laboratories; supervised faculty of 7.

Similarly, avoid long paragraphs that describe your work in detail. Employers who receive many resumes will not read long paragraphs. Instead, limit the resume to one page. Leave plenty of white space, use underlining or boldface, and lay out the page so that the important data are immediately apparent. If you have supporting information, like a long list of publications or specific job details, attach it on a separate page or pages.

I see no reason to include "personal data" such as hobbies or height and weight, though birth date and marital status might be appropriate in some resumes.

retrofit. Pure jargon, but a word for which I know no synonym. Use it!

retronym. Columnist William Safire coined this word, which is different from *back-formation*, to mean *a new word for an old thing*. *Mainframe computer*, for example, is a retronym: it refers to a device that would have been called simply a *computer* before the development of *minicomputers* and *microcomputers*. To distinguish the old from the new, the old had to be given a new name. If you use a retronym, however, make sure it is not just a **redundant expression**. See also **neologism**.

rhetorical question. Why not ask a rhetorical question? It can spark the reader's interest, as well as make your writing more informal and less stilted.

right-justified. *Justifying* is adjusting the edges of a column or line of type. Most typing, for example, is *left-justified*; that is, the left edge of the column defines a line down the page. The right edges of the lines are not aligned; this is usually called *ragged right*.

Many computers can *right-justify* a column of type; that is, the right edge of the column also defines a line down the page. Right-justified material looks much more presentable than ragged right. Unfortunately, however, many dot matrix printers justify the right edge of the lines simply by adding spaces within the line itself. This can make the material harder to read, so much so that many people disapprove of justifying the right edge when using a dot matrix printer.

Other printers have *proportional spacing*; these justify the right edge by spacing the letters evenly along the line. A right-justified text with proportional spacing is easier to read than a text without proportional spacing.

With either kind of spacing, you will have to hyphenate many long words if you choose right-justification. Otherwise, the type will be spread out unnecessarily and, again, the text will be harder to read. Unfortunately, hyphenating also makes the text harder to read.

For these reasons, many editors prefer typescripts to be printed ragged right rather than right-justified. Unless you are a professional printer, use right-justification only when overall appearance is more important to you than readability.

right word. Mark Twain said that the difference between the right word and the nearly right word is the difference between the lightning bug and the lightning. Work to get the right word. Get a **thesaurus** if necessary. Sleep on it for a few days. But don't settle for the wrong word or resort to a clumsy or invented term placed in quotation marks to tell the reader that you really knew better.

round. If you use it in place of *around*, don't apologize with an apostrophe, *'round. Round* is a perfectly good preposition, though it may be a bit too colloquial for scientific writing.

ruggedized. A terrible word that is really no different from *rugged* and makes your writing sound like **jargon**. If you need a verb, use *strengthened.*

rules. See Orwell's five rules under **barbarism**.

run-on sentence. This is really two sentences, they should be separated not by a comma but by a period or a **semicolon**. The sentence could, however, be rewritten properly as one sentence if you used a subordinate conjunction like *which*:

> The first sentence is really two sentences, which should be separated not by a comma but by a period or a semicolon.

S

sacrificed. In biology and medicine, often a **euphemism** for *killed*, so why not use *killed*?

same exact, same identical. Redundant synonyms for *same*. *Precisely the same* and *exactly the same* are also redundant, except perhaps in contrast with *approximately the same*.

sample, specimen. In statistics, a *sample* is a portion of a population; you study the sample to gain information about the population as a whole. A *specimen*, on the other hand, is usually a single member of the population; it may be a member of the sample as well. A specimen may not be representative of the population, whereas a sample is assumed to be representative.

In general usage, outside statistics, *sample* is synonymous with *specimen* or *example*. The distinction is a fine one and need not be made except in statistics or another context where *sample* could be ambiguous.

Incidentally, I think that, when the physician asks you for a specimen, she really means a sample.

saving, savings. *Savings* is not synonymous with *saving*, or, at least, it wasn't until Madison Avenue got hold of it. Not too many years ago, they started writing

> a savings of 30 percent

when they meant

> a saving of 30 percent.

What's the difference? Your life's *savings* are what you have accumulated during your lifetime; a *saving* is (or was) what you get by buying at a 30 percent discount. You don't, after all, write *a losses of 30 percent*, so don't write *a savings of 30 percent* either.

Incidentally, the correct locution is *Daylight Saving Time*, not *Daylight Savings Time*. Its opposite is *Daylight Wasting Time*, as anyone who has tried to play tennis in January will tell you.

scientific creationist. An *oxymoron*, or contradiction in terms. Creationism is not a science, because it sets out to prove a preconceived idea and does not accept evidence to the contrary. Science, by contrast, sets out to test hypotheses. However imperfectly it tests these hypotheses, it does not in principle begin with a preconceived idea. This is what sets science apart from pseudoscience.

seem. A lot of people write or say *It seems that* ... when they really should be making a flat statement. Don't say

> When there's smoke, it would seem that there might be fire.

See **would; vague words, vogue words**.

self-destruct. There is no verb *destruct*; the verb is *destroy*. I suppose, therefore, that you could say *self-destruct* is wrong. But, in fact, *self-destruct* is what people say and write, and I consider it perfectly acceptable.

semicolon. Sometimes two sentences are closely related in meaning, or one follows closely on the heels of the second; these sentences may be joined by a semicolon instead of separated by a period. Many good writers use semicolons to show this connection between two thoughts explicitly. See also **colon**.

Semicolons may also be used to separate the items in a **list**, especially if the list is long or some of the entries themselves contain commas. For example, in a reference list, you might write

> See papers by Young, Johnson, and Goldgraben; Franzen, Gallawa, and Day; and Cherin.

You could argue in favor of replacing the last semicolon by a comma, but I think the semicolon is better because it is more consistent and allows for no ambiguity.

Do not, however, use semicolons where commas are appropriate. Frequently, I see a semicolon used to set off an **appositive**:

> Photographic chemicals are no more toxic than other household chemicals; items like bleach, ammonia, and insecticide.

Here the semicolon should have been a comma or, perhaps better, a dash:

> Photographic chemicals are no more toxic than other household chemicals—items like bleach, ammonia, and insecticide.

Similarly, a semicolon should not be used to set off a nonrestrictive phrase:

> Some samples were thermally cycled; such as Samples A, B, and C.

This sentence would best be reorganized as

> Some samples, such as Samples A, B, and C, were thermally cycled.

sentence fragment. A short sentence that does not contain a verb is sometimes called a *sentence fragment*. *Why not?* is a sentence fragment, as is *Not true*. There is nothing wrong with a sentence fragment; many of the entries in this book begin with sentence fragments, and many **figure captions** are sentence fragments or begin with a sentence fragment. Sentence fragments are a good way to vary the length of your sentences and make your writing more interesting.

Some sentence fragments, like *Yipes!* are also exclamations. These should be followed by **exclamation points**, but a sentence fragment need not necessarily be ended with an exclamation point.

sentence length. One authority claims that those who advocate short sentences are not very bright and cannot understand complex material; they blame the style when, in reality, the material is too complex for them.

I do not agree with this analysis. Readability correlates with sentence length as well as with the difficulty of the material. Scientific and technical material is often inherently difficult. Your readers are intelligent, busy people, not poor readers demanding a primer or an "article in comic book form." Your job as a writer is to explain complex material as clearly as possible; you will do this best by writing simply and logically, and by presenting your thoughts sequentially. You can best present complex thoughts sequentially by isolating each thought and separating it from the others with a period or a semicolon. Short sentences will follow automatically.

But how short? Surely not artificially short, so that your writing is choppy. Nor all the same length, so that your writing is boring. But short enough that a single sentence does not cover more than a single thought or description.

Probably a good average to aim for is 20 to 30 or 35 words per sentence, on average. Obviously, some sentences will be shorter than 20 words, and some will be longer than 35 words. Textbooks and other lengthy or descriptive works should probably average nearer to 20 than 35, whereas short, dense articles or letters directed at specialists might average somewhat longer sentences. See **Flesch's readability index**.

sentence modifier. An adverb or adverbial phrase that usually precedes a sentence and modifies the whole sentence. See **adverb; dangling modifier**.

serves to. As in

> The copper layer serves to distribute the stress uniformly.

The phrase is meaningless; the sentence should be simply

> The copper layer distributes the stress uniformly.

sexist writing. See **he or she**.

Shakespeare. Flesch is not alone when he asks, Did Shakespeare make mistakes? Many people point to the errors made by the man they consider to be our greatest writer and argue that they must not be errors if Shakespeare wrote them. I will not contest the point, except to note that Shakespeare may not be the best model to use for technical writing, or for modern writing of any kind. Shakespeare, after all, wrote *methinks*, which was correct then but not even cute today. The language changes every day, and I look to modern writers for instruction on modern writing.

shall. The military uses it for *must*. In everyday U.S. English, there is no difference between *shall* and *will*, and, in reality, Americans almost never use *shall* (though the British do use *shall* and *will* differently). Ignore that prescriptivist stuff you learned in high school and write *will* when it seems natural.

share with. Use when you literally mean to share something. Do not use as a synonym for *tell.*

she, s/he. See **he or she**.

shook. The past participle of *shake* is *shaken*. Except perhaps in the idiomatic *shook up* for *distraught*, the correct form is *shaken*:

> The scientific world was shaken up by the discovery.

shortlist. A barbarism meaning something like *narrow down to*:

> The Department of Energy has shortlisted three sites.

It can also be used as a noun, which really looks like jargon. What is the matter with *list* or, if you insist, *short list*?

SI units. Most of the sciences use the metric system of units. Present-day metric units are known as *SI units*, from the abbreviation for the French *Système International.* Decisions concerning SI units are made by several international committees and are published periodically by the International Bureau of Weights and Measures, or BIPM (Bureau International des Poids et Mesures). Here, we are concerned not with the definitions of the quantities, but with the use of SI units in writing.

Write the units in Roman type, not italic. (In this entry, I have departed from my usual habit of writing examples in italics, so that the SI abbreviations will appear in Roman type.) When you spell the name of an SI unit fully—hertz, decibel, mole, meter—do not capitalize it, even if the unit is named after a person. (At the beginning of a sentence, in a title, or in a table, treat the name of a unit like any other word.)

Form the plurals of SI units as you form any other plurals—ten meters, 1.3 moles, 3 decibels. (Many people find it hard to say *a frequency of 100 megahertzes*, but *100 megahertz* is arguably incorrect. We say, after all, *two klutzes*; why not *two hertzes*?)

Quantities less than 1 are singular: 0.5 meter, not 0.5 meters; 1/10 joule per pulse, not 1/10 joules per pulse.

When you abbreviate a unit—Hz, dB, mol, m—capitalize the abbreviation only when the unit is named after a person. (Don't ask me why this seeming inconsistency.) Do not end the abbreviation with a period and do not make it plural with an *s*.

The International System also calls for a space between a number and a unit or its abbreviation. That is, write 9.8 cm, not 9.8cm. This rule does not imply to me that you may not insert a **hyphen** in the space when style requires it; that is, you may write either 1 km fiber or 1-km fiber, depending on your hyphenation style. When I have my choice, I will omit the hyphen, since there is rarely an ambiguity. See also **numbers in sentence**.

Do not capitalize metric prefixes, such as mega-, micro-, kilo-, when you spell them fully. Capitalize the prefixes that represent multipliers larger than 1000 only in their abbreviated forms, M, T, G. The abbreviation for kilo-, however, is always a lower case *k*, presumably to avoid confusion with K for the kelvin. In addition, the abbreviation for the liter may be capitalized to avoid confusion with the numeral 1: 10 l or 10 L. In the U.S., the capital L is preferred. (Since a liter is a cubic decimeter, some writers use dm^3 as the abbreviation for the liter. Before 1964, the liter and the cubic decimeter differed slightly. Perhaps for this reason, the liter is not recommended for reporting the results of precise volume measurements. The decimeter is a very odd unit; except for reporting precise volume measurements, stick with the liter.)

We used to write 0.633 millimicrons, but SI prohibits using compound prefixes. The SI base unit for mass is the kilogram, but do not write millikilogram for gram, nor kilokilogram for megagram. Similarly, do not write 1.2 decimillimeters for 0.12 millimeter or 120 micrometers.

Do not use parentheses when forming powers of SI units: MHz^{-1}, cm^3, not $(MHz)^{-1}$, $(cm)^3$.

Some of the names and abbreviations are different in SI than they were in the past. In particular, use micrometers, not microns; s, not sec; g, not gm; and K, not °K. (Do not, however, omit the degree sign with degrees Celsius: 273°C, not 273 C, nor 273 °C. See **degree**.)

Although it is correct to spell the names of the units—1 watt per square centimeter—full spellings and abbreviations should not be mixed (or, as I like to say, do not mix complete spellings and abbrs.). Therefore, 1 watt/cm^2 should be changed to 1 W/cm^2, and 1 watt/square centimeter should be changed to 1 watt per square centimeter. It is not, however, necessary to write one watt per square centimeter; Arabic numerals and complete spellings mix fine. Avoid, however, one Hz or a sentence like

> To cool an atom from 300 K to mK temperatures requires 10 000 scattering events.

Instead, rewrite the sentence to use complete spelling of the unit when it is not associated with a number:

> To cool an atom from 300 K to millikelvin temperatures requires 10 000 scattering events.

See also **numbers in sentence**.

The use of the slash to indicate a fraction can sometimes cause ambiguity. For example, does 1 J/cm^2/s mean 1 J · s/cm^2 or 1 J/(cm^2 · s)? Often the answer is obvious from context or to the specialist, but repetition of the slash in this way should be avoided. Instead, write the units with a single slash, 1 J/(cm^2 · s), or with negative superscripts, 1 J · cm^{-2} · s^{-1}.

In the U.S., the centered dot is preferred as the symbol for multiplication of units: 10 N · m, for example. Internationally, however, two other forms are also allowed: 10 N . m or 10 N m. When spelling the units, do not use the centered dot, but rather a space or, alternatively, a hyphen: 10 newton meters, or 10 newton-meters. Similarly, do not use a slash, 10 meters/second, but rather use the word *per*: 10 meters per second.

Certain other units, such as the day, degree, and metric ton, are considered "in use with" the International System. In addition, there is no prohibition against mixing SI units with certain quantities, usually dimensionless, for which there is no SI equivalent. For example, how else would you write 100 Gbit/s or 10 joules per pulse? (Do not write, incidentally, 100 Gbits/s; that is, do not make Gbit plural with an *s*.) Do not, however, mix SI units and units that are explicitly not part of the International System. See also **CGS units**.

TABLE SI-1. Some SI units and their symbols.

Quantity	Name	Symbol
length	meter	m
mass	kilogram	kg
time	second	s
electric current	ampere	A
temperature	kelvin	K
amount of substance	mole	mol
luminous intensity	candela	cd
plane angle	radian	rad
solid angle	steradian	sr
frequency	hertz	Hz
force	newton	N
pressure	pascal	Pa
energy, quantity of heat	joule	J
power	watt	W
electric charge	coulomb	C
electric potential	volt	V
capacitance	farad	F
resistance	ohm	Ω
conductance	siemens	S
magnetic flux	weber	Wb
magnetic flux density	tesla	T
inductance	henry	H
luminous flux	lumen	lm
illuminance	lux	lx
activity (of a radionuclide)	becquerel	Bq
absorbed dose	gray	Gy
dose equivalent	sievert	Sv
gain or loss	decibel	dB
phase	neper	Np
hydrogen potential	—	pH

TABLE SI-2. SI prefixes.

Factor[a]	Prefix	Symbol	Factor[a]	Prefix	Symbol
18	exa	E	−1	deci	d
15	peta	P	−2	centi	c
12	tera	T	−3	milli	m
9	giga	G	−6	micro	μ
6	mega	M	−9	nano	n
3	kilo	k	−12	pico	p
2	hecto	h	−15	femto	f
1	deka	da	−18	atto	a

[a] Exponent of 10; that is, 18 means 10^{18}

TABLE SI-3. Other acceptable units.

Name	Symbol	Name	Symbol
minute	min	degree	°
hour	h	minute (of arc)	′
day	d	second (of arc)	″
liter	L, l	electronvolt	eV
hectare	ha	unified atomic mass unit	u
metric ton (tonne)	t		

TABLE SI-4. Obsolete units (other than English units).

Name	Symbol	Name	Symbol
millimeter of Hg	mmHg	rad[a]	rad
torr	Torr	rem[a]	rem
bar[ab]	bar	roentgen[a]	R
atmosphere	atm	curie[a]	Ci
kilowatt-hour	kWh	calorie	cal
barn[a]	b	gal[ab]	Gal
kilogram force	kgf	angstrom[ab]	Å

[a] Temporarily accepted by BIPM.
[b] Discouraged by American National Standards Institute.

sign off. Some managers do not simply sign forms; they *sign off on* them. The *on* cancels the *off*, and they end up signing them after all.

signal-to-noise ratio. This should not be shortened to *signal to noise*, without the noun *ratio*. Common abbreviations are *SNR, S/N*, or, less commonly, *S/N ratio*. See also **hyphen; slash**.

since. See **because, since**.

singular or plural. Sometimes it is not obvious whether to use a singular or a plural verb. The purist says to use the singular verb if the subject is grammatically singular. But we say

> There was a lot of milk,

and

> There were a lot of jelly beans.

Here, *lot of* and *lots of* have become almost like adjectives in that *lot* and *lots* have lost their meanings as nouns. The verb agrees with the object of the preposition, in this case, either *milk* or *jelly beans*. The phrases, *lot of* and *lots of*, function as units to modify the nouns. So use whatever verb sounds right to you when you use an expression like *a lot of*.

Likewise,

> Twenty percent of them were surveyed

sounds all right, even though you could argue that *percent* is singular. You might, though, write

> One percent of them was surveyed

because of the influence of the word *one*.

Proximity, in other words, has a lot to do with choosing the verb. You might write

> One percent of the population was infected,

but

> One percent of the men were infected.

The closeness or proximity of the plural *men* somehow calls for the plural verb. The prescriptivist would insist on the singular verb, but I would defend the plural here.

Sometimes a pair of nouns sounds so good together that we forget to use a plural verb:

> Bias and prejudice exists in northern cities as well

may not be good, but it is at least defensible if we think of *bias and prejudice* as a **collective noun**. I prefer, however, to limit such constructions to speech and avoid them in writing.

Number of and *none of* can be a lot like *lot of*. Do you say

> A number of methods *has or have* been developed?

In writing, I stick with the singular, because the purist in me sees the singular *number* and thinks the verb cries out for agreement. But many, if not most, native speakers think of *a number of methods* as a collective noun and use the plural verb. Apparently, they hear the plural *methods* rather than the singular *number* and want the verb to agree with the object of the preposition (which is closer to the verb). Recently someone said to me,

> None of the clarinets *were* able to do it, but none of the band *was* willing to help them.

Each of them is another example of a phrase that should grammatically take a singular verb, but many speakers would say *each of them were* . . . . Here I side with the prescriptivist: *each of* does not function like *lots of* and sounds singular to me.

Many experts have already thrown in the towel on this point and accept the plural verb, even though logically the singular verb is called for.

The number of (as opposed to *a number of*), on the other hand, is very definite and must take a singular verb:

> The number of insect species exceeds all others combined.

situation. A great word for the jargonauts! I recently heard two examples that use *situation* completely superfluously:

> In a jeopardy situation

and

> How's your productivity situation for those?

See also **dead words; vague words, vogue words**.

slang. A *slang* word is one that has spread from a narrow group to the general population. The line between slang and colloquial or informal speech is hard to draw; you just have to feel it. *Cool* and *groovy* are slang, and *awesome* has become slang, but what about *pothead* or *uptight*? I can imagine wanting to use such words in certain kinds of writing, and their use will only improve your writing by making it simpler and more accessible. For example, if I were writing for a science news magazine, I would not write *marijuana user* or *person addicted to cannabis* over and over when I could write *pothead* instead. I would use a word like *pothead* in a technical conference proceedings or in a talk, but I would probably forgo it in an archival journal, especially if I thought the word was likely to die out. Similarly, in formal writing, I would not use slang expressions like *op-amp* (for *operational amplifier*), *lab*, or *gas* (for *gasoline*).

Avoid slang in most technical writing, but do not be afraid of colloquial or newly coined phrases where they are appropriate. See also **jargon; neologism.**

slash. Also called *virgule, solidus,* or *diagonal stroke, the slash* (/) used to be reserved for division or as a symbol for *or*, before Madison Avenue got hold of it. I am going to lose this one, but I argue that it is improperly used in place of a hyphen and elsewhere, because that dilutes its unique meanings. For example, *300 watt/seconds* should have been *300 watt-seconds* and could have been confusing if the reader misread *watts per second.* A graph axis labeled *Frequency/MHz* should have been *Frequency, MHz* or *Frequency (MHz)*, so that the label would not look like *Frequency per megahertz.* Even my copy of Strunk and White is labeled *The Elements of Style 3/e*, presumably for *third edition.*

Technical writers also use the slash when they have too many hyphens to contend with, as in *silicon-oxide/silicon-dioxide layer.* This would be clearer as *a layer of silicon oxide and silicon dioxide* or *a layer of silicon oxide on silicon dioxide*, whichever is meant. The shortest way is not always the best.

Sometimes the slash is used in a way that may be ambiguous, if only because it is continually misused in place of a hyphen. For example, are *yellow/green spots on a red background* spots that are yellowish-green, or are they spots that may be either yellow or green? Likewise, in a photography magazine, I ran across a table headed

CONTRAST
50 mm/30 lines/mm.

It turned out to mean *the contrast of a 50 millimeter lens measured at 30 lines per millimeter*, but overuse of the slash made it ambiguous if not unintelligible.

slides, viewgraphs. Kodak says that a typical screen with dimensions 1.3 × 2 meters may be viewed from 20 meters away. This is the equivalent of viewing a 24 × 36 millimeter slide from about 35 centimeters, almost 50 percent farther than a typical reading distance. The screen has to be legible to a listener seated 20 meters from the screen, so the slide has to be legible from 35 centimeters. If you want to put text on the slide, you will have to limit the text to what you can read from that distance: no more than nine lines of doubly spaced text.

Typing is a little light for slides, but it is convenient. If you want to make a slide by typing the text, Kodak says to draw a rectangle about 11.5 × 7.5 centimeters and use it as a template.

If you prefer hand-lettered or computer-generated characters, or characters generated by a lettering machine, this comes to 7 millimeter high letters, or 24 point letters, drawn on a standard 22 × 28 centimeter sheet of paper, oriented horizontally. (A typesize of 72 *points* corresponds to capital letters about 19 millimeters or 0.75 inch high.) Letters on the screen have to be about 25 millimeters high for every 10 meters of viewing distance.

The letters should be drawn with about 1 millimeter lines; this corresponds to a typeface known as Helvetica medium, for example. Capital and lowercase letters are usually more readable than all capitals, and I prefer *sans serif* letters, or letters that do not have little ornaments, called *serifs*, like most newspaper and magazine printing. Use a broad felt-tipped pen or a drawing pen with about a 1 millimeter point, if you are lettering by hand, and print neatly. Leave plenty of space, at least the width of a lowercase *n*, between words, and twice that between sentences.

Likewise, leave at least the height of a capital letter between lines.

Kodak points out that legibility in one form does not guarantee legibility in another. If your slide is to be legible in the back of the room, the 22 × 28 centimeter original will have to be legible to you from about 3 meters away if it is to be legible to the listener who sits 20 meters from the screen. If you use an ordinary ball-point pen, the text will not be legible from those distances.

If you make viewgraphs (22 × 28 centimeter transparencies intended for an overhead projector), follow the guidelines for slides lettered on 22 × 28 centimeter paper. If you orient the viewgraph vertically and want the whole page to appear on the screen, you might want to draw your letters about 30 percent larger, that is, 9 or 10 millimeter, or 32 point, letters.

If you are teaching a course in which everyone gets a copy of the slides, do not assume that your slides can be illegible on the screen just because they are legible in the notebook.

A good way to make viewgraphs is to draw on the back of a sheet of graph paper. Photocopy the drawing onto transparency material. If you are careful, the grid will not copy. If you expect to use the same viewgraphs many times, consider placing each one inside a transparent plastic sleeve. These may be purchased with loose-leaf holes and sometimes come with a sheet of heavy black drawing paper inside. If you discard the paper or cut a frame out of it, you can protect your viewgraphs, as well as write over them without destroying them.

Keep **figures; graphs;** and **tables** simple too. In particular, do not try to reproduce a complicated drawing with tiny letters; see whether you can read the drawing from the appropriate distance before you decide to make a slide or viewgraph. Figures copied from journals are often acceptable because they are drawn with heavy, black letters, whereas many textbook figures are too complex and have tiny letters. Drawings may be fine in the textbook because they can be studied at leisure, but they may be poor on the screen, because they may disappear in less than a minute. Similarly, consider replacing a table with a bar graph or a circle graph (pie chart); though less precise, the graph can be understood more easily in the short time it is on the screen.

so. Colloquially, the adverb *so* can be a synonym for *very*, as in the exclamation, *It was so good!* But in writing, it is probably best to use *very* when you mean *very*.

So is most commonly used as a conjunction:

> They wanted to drill a hole, so they aimed the laser at the wall.

It should not be preceded by *and*:

> They wanted to drill a hole, and so they aimed the laser at the wall,

and should not, in general, be used to begin a sentence. In New York, however, most spoken questions seem to begin with *so* (*so when did ya get here?*).

Some people consider *so* to be synonymous with the conjunctive adverb *therefore*; so they argue that it should be preceded by a semicolon. This usage may be elegant, but it is also archaic. More important, many will consider it downright wrong, so I avoid it.

so as to. To.

so-called. Use *so-called* only when you want to denigrate something, as in calling a pseudoscientist or a creationist a *so-called scientist*. Do not, however, use *so-called astrologer*, since that is exactly what astrologers are called. Likewise, if you are introducing a new device, do not call it *a so-called X-ray laser* when an X-ray laser is exactly what it is. In the sentence

> A comparable pattern of divorce patterns is not available for so-called primitive cultures,

however, *so-called* is correct because the author was calling into question the adjective *primitive* as applied to less-technological cultures. Unless she was trying to make a point (nothing wrong with that!), though, she should probably have chosen a better word than *primitive*.

so that, such that. Most of the time, *such that* should be replaced by *so that*.

> The joint was oriented such that the field was perpendicular to the wire

means

> The joint was oriented in such a way that the field was perpendicular to the wire

and should have been written

> The joint was oriented so that the field was perpendicular to the wire.

Whenever *in such a way that* works, use *so that*, not *such that.*

Such may also be used as an adjective, not an adverb, and should be used for comparison or specification of a noun. *Such that* is usually better broken up. We say

> It was such a powerful storm that it knocked over a telephone pole,

not

> It was a powerful storm, such that it knocked over a telephone pole.

Likewise, in technical writing, break up *such that* for more natural word order and easier reading.

So is often superfluous when connected to *that.* In

> The telescope is sufficiently stable so that the images hold together,

delete *so*:

> The telescope is sufficiently stable that the images hold together.

(See, incidentally, **sufficiently**.)

So that would be correct, on the other hand, if the intended meaning were something like *with the (intended) result that*:

> The telescope had to be stable, so that the images would hold togehter.

some. Not a good synonym for *about* or *approximately.*

> Some 45 papers were presented at the conference

would be better as

About 45 papers were presented at the conference.

See also **about; as many as; roughly**.

some time, sometime. *Sometime* is an adverb and refers to an unspecified time in the past or future. It is accented on the first syllable, like most such compounds. *Some time*, on the other hand, is a noun phrase and refers to a somewhat indefinite but usually significant period of time. It is accented on the second syllable. The pronunciation gives a clue to which form is correct.

Sometimes (an adverb that means *occasionally*) and some time both may be grammatically correct, but using the wrong form may give the wrong meaning. For example,

> One problem was the demise of the custom sometime before the researcher took up residence.

As it stands, this sentence means that the custom ended *at some unspecified time* before the researcher arrived. Possibly that is what the author intended. It seems at least as likely to me, though, that the author meant

> One problem was the demise of the custom some time before the researcher took up residence;

that is, the custom had been extinct for quite a while before the researcher arrived.

somewhat of. *Somewhat* is usually used as an adverb meaning *to some extent*. *Somewhat of a vague standard* is more properly rewritten *a somewhat vague standard.* See also **big of**.

sort of. See **kind of, sort of**.

space. When I hear someone say *I am in a bad space today*, I wonder whether he lives in some multidimensional universe I do not have access to. (A pseudopsychologist would say he does.) This is a barbarism akin to *I know where you are coming from* and *I can't relate to that.*

On a more serious note, always leave a space between a number and a unit: 5 J or 5 joules. Indicate multiplication either by leaving no space, 5*J*, or by using a raised dot, $5 \cdot J$. Then, even in a typescript, the product 5*J* cannot be mistaken for 5 joules. See also **SI units**.

space of time. Only in relativity are time and space related. Use **duration** or *period*, preferably without the superfluous modifier *time*. See also **point in time; time frame**.

speak to that point. You don't speak to a point; you speak *to* someone and *about* a point.

speeches and talks. See **conference proceedings**.

special order. When you need the verb, use *to order specially*, at least in writing.

split infinitive. This is a purely artificial prohibition that derives from the time when languages were studied by using Latin grammar as the paradigm. In reality, there is only one reason to not split an infinitive: so that no one will think you don't know any better. See **permissivist, prescriptivist**.

To avoid splitting the infinitive, many resort to putting an adverb right before it:

> We wanted the values wholly to be included within the range.

It sounds less pompous, especially with such a *compound infinitive*, to write

> We wanted the values to be wholly included within the range.

You could also write

> We wanted the values to be included wholly within the range,

but that might cause a slight shift of emphasis.

> Just to maintain saturation, we must reduce the pressure

is ambiguous; if you mean

> To just maintain saturation, we must reduce the pressure,

write it that way.

square brackets. See **brackets**.

stock phrases. It would take too long to write a short **business letter**, for example, without using stock phrases, and there is no reason to avoid stock phrases. Just don't let them look like stock phrases. Instead of the very formal, not to say pompous, *Enclosed please find* . . . , write something like *Here is* . . . . Let *your* stock phrases be short and to the point.

Likewise, in a technical paper, you do not need phrases like *The purpose of this paper is to* . . . , nor, at the other end of the paper, *In conclusion*, ... Just say what you have to say, write your conclusions, and stop.

straight line. OK in nontechnical speech, but in the technical world all lines are straight.

straightforward. One word.

style manual. Most publications use a *style manual*, a book or booklet that tells how the editors interpret certain more-or-less optional rules such as hyphenation, reference format, or use of commas. When you are writing for these publications, become familiar with their style manuals. Using the style manual is not just a courtesy to the editor; it also ensures that your paper will suffer less editing and therefore will be less prone to accidental errors. If you cannot find the style manual, examine the journal or publication medium until you know how they operate.

When editors change your manuscript "for reasons of style," incidentally, that does not necessarily mean that there was anything wrong with your style, nor, as some have inferred, that they think their ear is better than yours. It merely means that you did not conform to their rules.

subjunctive. The *subjunctive mood* has not quite disappeared from English. It still persists in such phrases as *as it were* and *God forbid!* and in clauses like *If I were you.* When the professor says

I recommend that everyone be ready for the test,

he is using the subjunctive mood. So am I when I say

It would be inconsistent if I *disbelieved* in causality,

even though I am writing about my present belief.

What is funny about these constructions is that the form of the verb is different from that in the ordinary, *indicative* mood. What is funnier is that not one of them sounds strange. The subjunctive is used, often quite naturally, when we express doubt, hope, or a condition contrary to fact.

Sometimes the subjunctive mood sounds pompous, especially when the present subjunctive is called for:

> I doubt that the apparent randomness at the submicroscopic scale be real, rather than apparent

may be correct, but I would change *be* to *is.*

Moral? Don't worry about the subjunctive mood. Just write what sounds most natural.

such as. Use *such as* when the list that follows it is incomplete. Do not end the list with **etc.**, since *such as* implies *etc.*

> The data include errors such as scattering, nonlinearity, noise, etc.

should be rewritten as

> The data include errors such as scattering, nonlinearity, and noise.

See also **etc.; include; like; semicolon**.

such that. See **so that, such that**.

sufficiently. A long word meaning nothing more than *enough.*

> The pressure was sufficiently high to suppress cavitation

should be rewritten

> The pressure was high enough to suppress cavitation.

sulfur. In the U.S., the preferred spelling.

super, ultra. In a day when reasonably proficient players are called *superstars*, can you blame a scientist for developing *a supercold* *refrigerator* or *an ultrahard coating*, or cooling *to an ultralow temperature such that the deBroglie wavelength is long compared to atomic dimensions*? I suppose not, but, still, these prefixes are best avoided. Do not overstate your case. Not only do you risk overselling your paper, but you also add to the inflation of the language. See also **vague words, vogue words**.

surrounded on three sides. Nearly surrounded.

surveille. A back-formation from *surveillance*, this barbarism should be replaced with *put under surveillance*, *put under observation*, or *watch*.

syllabification. Do not hyphenate a word at the end of a line arbitrarily; check a **dictionary** and break the word between two syllables. If a word has an ending, such as *-ing*, break the word before the ending: *sort-ing*, not *sor-ting*. See also **right justification**.

symbols. In archival publications, mathematical variables are usually italicized; in camera-ready documents prepared from typescripts, they are apt not to be. Nevertheless, it is customary, when using symbols in sentences, not to set them off in commas:

capacitance C per unit length,

rather than

capacitance, C, per unit length.

This word order, incidentally, seems better to me than

capacitance per unit length, C,

since the symbol C is in apposition to *capacitance*, not *length*.

When you want to show multiplication, do not leave a space between two symbols or between a number and a symbol: use instead $15g$ or $15 \cdot g$, for example. Especially in a typescript, where you might not have italics, 15 g, with the space, might be mistaken for 15 grams. Incidentally, if g is the acceleration due to gravity, write $15g$, not 15 g's.

Unlike mathematical variables, the abbreviations for **SI units** should not be italicized: 440 Hz, not 440 *Hz*. See also **numbers in sentence**.

T

table. Use tables when you want to convey a fairly large quantity of precise information, more precisely, for example, than you can convey it in a **graph**. In general, number tables consecutively, in the order in which they are mentioned in the text. Even if you have only one table, most publications ask you to call it *Table 1*, rather than just *the table.*

Give each table a **title**, and label each row and column clearly. If physical or other units are associated with any of the entries, they are usually best placed with the row or column label: *Mass, kilograms* or *Number of samples (thousands)*. Here is an example of a short table.

TABLE 1. Specifications of typical microscope objectives.

Power	Focal length,[a] mm	Numerical aperture	Useful magnifying power[b]
10 ×	16	0.25	75 ×
20 ×	8	0.45	150 ×
40 ×	4	0.65	180 ×
100 ×	1.6[c]	1.3	400 ×

[a] Assuming 160 mm tube length.
[b] Approximately 300 × *NA*.
[c] Oil immersion.

This table has a descriptive title. There is a rule between the title and the column headings, and between the column headings and the rows. There are extra spaces both below and above the table. The rows are widely spaced, and there is not so much information that it is squeezed and therefore bewildering. (If, like me, you can't distinguish between the rows and the columns, just remember that the columns are the vertical entries, because columns stand upright.) In addition, the decimal points, whether explicit or implicit, are aligned vertically, and there are no **naked decimal points**.

Footnotes that relate specifically to the table are indicated with symbols (here letters) not being used to flag **references** in the text; they are placed directly at the bottom of the table, not the bottom of the page. Do not use superscript numbers in the table if you are using such numbers in the body of the text to indicate references. If you are using square brackets in the text for reference citations, then superscript numbers are appropriate in the table. In general, however, do not put a separate list of references under the table, unless you need, for example, to cite several sources of the data.

task force. This term borrowed from the military usually means no more than *committee*. It is used by weak leaders to give the appearance of firmness and activity.

technocrat's law. Volunteer no information, but do not lie in response to a direct question. This is closely related to the *zeroth commandment*, cover thy rear.

tee, vee. Anything shaped like the letters T or V; a tee connector, for example. You need not enclose these perfectly good English words in quotation marks, nor italicize or capitalize them. (This entry does not imply that I am a TV addict.)

telecon. A bureaucrat's word for *telephone conversation*.

tend to. Usually unnecessary. In

> Many procedures tend to introduce stress in the fiber,

delete *tend to*.

tense. What tense do you use in a technical paper? That depends on what you are saying. In general, use the present tense, except where you are obviously referring to something that happened in the past.

State universal facts in the present tense:

> Water freezes at 0° Celsius.

On the other hand, state your observations in the past tense:

> The solution boiled at 93° Celsius.

Graphs, tables, and references exist now. Refer to them in the present tense: *Table 2 shows that . . .*; *Reference 7 describes. . . .* On the other hand, *Geronovitch showed in Ref. 7 . . .* is all right, since Geronovitch's work is in the past.

In a nutshell, *Data were taken . . .*, but *The data show that. . . .*

Excessive use of present **participles**, especially *having*, in technical writing leads to tense inconsistencies. For example, the sentence,

> Years ago, we used pumps having a flow rate of
> 10 L/s,

is inconsistent because the verb *used* is in the past tense, but the participle *having* is in a present participle. The sentence will be consistent if rewritten

> Years ago, we used pumps that had a flow rate of
> 10 L/s.

(I confess to exaggerating the inconsistency with the words *years ago*; still, the subordinate (*that*) clause is better style than the participial phrase.)

term-ite. There are lots of term-ites around, and they come out of the woodwork every few years to complicate matters by changing *the millimeter of mercury* to *the torr* and *the cycle per second* to the less descriptive *hertz*. Now they want to give a name to the *newton-second.*

test procedure. Usually, *test* will suffice:

> We designed a test procedure to measure the
> attenuation

is no better than

> We designed a test to measure the attenuation.

than. *Than* is a conjunction and usually takes the nominative case. In speech this may not be so important; in writing, however, take care to use the right case after *than*. There is a big difference between

> He likes her better than I [do]

and

> He likes her better than [he likes] me.

Many writers use **compared to** or *compared with* and sometimes **relative to** when they mean *than*.

Do not, incidentally, use the incorrect spelling *then* for *than*.

that (conjunction). For clarity, don't forget the conjunction *that* after a verb like *show*, which takes a direct object.

> He proved the principle, a conjecture, really, almost
> as old as the field itself, was untrue.

This sentence lacks the subordinate conjunction *that*. Although it is not incorrect, I might, on cursory reading, miss the verb and think he had proved the principle. But he didn't, and if the writer had used the subordinate conjunction, you would have immediately been cued to look for a verb:

> He proved that the principle, a conjecture, really,
> almost as old as the field itself, was untrue.

Especially when a sentence is very long, use *that* to tip off the reader that a clause, not just a noun, follows.

that, which. Many technical journals make the prescriptivist's distinction between *that* and *which*. This distinction seems artificial to me, and it may originally have been based on an incorrect inference. But here it is. *That* is used to introduce a *restrictive clause*, that is, one that is essential to the meaning of the sentence:

> The gene that causes luminescence was spliced into the
> chromosomes of a plant.

Here I do not know what gene is meant without the *that* clause.

Which, on the other hand, is reserved for *nonrestrictive clauses*, those that are not essential to the meaning.

> This gene, which was isolated only recently, was
> spliced into the chromosomes of a plant.

I can understand this sentence without the *which* clause, which only gives additional information about the gene we have been talking about and is therefore parenthetical.

In plainer English, to the prescriptivist, *which* clauses are those that you would naturally set off by commas, and *that* clauses are not.

Usually, however, U.S. speakers use *which* for both restrictive clauses and nonrestrictive, but find *that* clauses uncomfortable when set off in commas. (We always say *in which*, *of which*, and *that which*, whether the clause is restrictive or nonrestrictive.) *That* sometimes sounds wrong; then, replace it by *which*. But watch out! Some *which doctor* will probably come along and change it.

Sometimes the relative pronoun is superfluous and should be removed:

> The term which is related to the period of oscillation was calculated

would benefit from removal of *which is*. It would be even better in the **active voice**:

> We calculated the term related to the period of oscillation.

that is. Usually a nonrestrictive phrase (see **that, which**) and therefore set off by commas:

> Our model will yield a complete description, that is, a description including both amplitude and phase.

Since the sentence can be understood without the phrase *that is*, the phrase is nonrestrictive. Many writers forget the second comma. Don't; either a phrase is set off with commas or it is not.

Sometimes *that is* introduces an independent clause. Then it should be preceded by a semicolon:

> Our model will yield a complete description; that is, it will yield a description including both amplitude and phase.

It is still followed by a comma, since it is nonrestrictive. See also **that** (conjunction).

that which. Often can be replaced with the simpler and less formal *what*. For example,

> The intensity is less than that which would be observed with a closed path

would be slightly easier to read written

> The intensity is less than what would be observed with a closed path.

the. A true purist would say that you should write

> the wavelength of 500 nm,

not

> a wavelength of 500 nm,

since there is only one wavelength of 500 nm. All other wavelengths are not 500 nm. I have used this construction, but it now seems forced to me.

Using *the* before a symbol is not good style, however. Do not write

> The E was in the direction of propagation

or

> The E was equal to 10 V/m.

Instead, use

> The electric field vector was oriented in the direction of propagation,

> E was parallel to the direction of propagation,

or

> The value of E was 10 V/m.

An exception: When you have a series of terms, $C_1, C_2, \ldots, C_n$, writing

> The C's are the coefficients in a power series

seems OK to me. But using *the T* as shorthand for *the temperature T* is very clumsy.

the author, the experimenter, the writer. Clumsy substitutes for *I* or *we*. See **first person**.

the present. As in *the present paper, the present work*. These are clumsy ways of saying *this paper* or *this work*. For example,

> The apparatus used in the present series of experiments is shown in Fig. 3

ought to have been written

> Our apparatus is shown in Fig. 3.

The present moment and *the present point in time* are barbarisms for *now*.

the reader. Don't be afraid to write **you** when you think it is appropriate. Calling the reader *the reader* is usually as clumsy as calling yourself *the writer*.

their. See **he or she; everybody, everyone**.

then. An adverb, and very clumsy when used as an adjective, as in *the then president*. If you have to specify use *who was then president*; if it is obvious, leave the qualification out altogether. Another example,

> They knew the by-then ancient mythology of a land of plenty,

would be much less clumsy as

> They knew what was even then the ancient mythology of a land of plenty.

See also **above, below; now**.

thence, whence. *Thence* is an obsolete word. It is not a synonym for *then*, but I have seen it used thus in manuscripts. *Thence* is closer to *thereafter* than it is to *then*. It should not be used in a sentence like

> We pumped the system out and thence baked it.

Unless you continued to bake it out indefinitely, rewrite the sentence

> We pumped the system out and then baked it.

Whence does not mean *where*, but rather *from where*. *From whence* is incorrect; *whence* is merely archaic and a little pompous, which aren't nearly as bad as incorrect.

there. There is no reason not to begin a sentence with *there*, but sometimes technical writers use this construction because of an apparent similarity to the **passive voice**.

> There exists a kink in the curve in Fig. 4

is unnecessarily convoluted.

> The curve in Fig. 4 shows a kink

is clearer and more to the point. Similarly,

> There were found to be anomalies in the data

would be better rewritten

> Anomalies were found in the data

or, if you want to take credit for your work,

> We found anomalies in the data.

There is also used by authors given to **circumlocution**. For example,

> There is a similar remark of Michelson's that is often quoted

is a roundabout way to write

> A similar remark of Michelson's is often quoted.

See also **it**.

thermal heating. All heating is thermal. This redundant phrase is sometimes used incorrectly to distinguish from *radiative heating*. Probably the author meant *conductive heating*.

thesaurus. A book or dictionary that lists synonyms, antonyms, and related words. Besides the familiar *Roget's Thesaurus*, I like *Webster's Collegiate Thesaurus*. *Roget's Thesaurus* groups words that have similar meanings in one or two lists; to find a word, you have to use the index to find the right listing and then search for your word among its relatives. The job is sometimes overwhelming, but *Roget's Thesaurus* is comprehensive. *Webster's Collegiate Thesaurus*, by contrast, lists words alphabetically, so you can immediately spot your word and its synonyms. Like *Roget's*, *Webster's* also supplies near-synonyms and antonyms.

Use a thesaurus when you can't quite find the **right word**. But use it with some care: One of the dangers of using a thesaurus is the opportunity to use a rare or inappropriate word. Unless you have a very good reason to do otherwise, use a word only if you would recognize it immediately. It is pompous to use a long or unusual word when a short, common one will do.

they. Often used for everybody or, rather, for people in general, without antecedent.

> They argue that evolution is established fact

means that scientists in general argue that way. Be sure, though, not to mix the *number* of your pronouns:

> If one worries about toxic vapors, they [or you] should use a fume hood.

Rather, change to

> If you worry about toxic vapors, use a fume hood.

See also **everybody, everyone; it; one**.

this. In the sentence,

> The surface of the glass has many microscopic scratches; this causes it to fracture well below the intrinsic strength,

you may well ask, To what precisely does *this* refer? It refers to the entire clause before it: to the statement that the glass has microscopic scratches.

Although still sometimes denigrated, this construction is often perfectly acceptable. In fact, it is much stronger than a similar construction that uses a participial phrase:

> The surface of the glass has many microscopic scratches, causing it to fracture well below the intrinsic strength.

The participial phrase is weak here, as is a *which* clause,

> The surface of the glass has many microscopic fractures, which causes it to fracture well below the intrinsic strength.

The last two constructions arguably display **misplaced modifiers**, because neither the participle *causing* nor the *which* clause is intended to modify the noun *scratches* that directly precedes them. (*This* does not introduce a modifier, so the first construction has no danger of dangling or misplacing a modifier.) More important, neither of the last two constructions makes clear that it is *the condition of having scratches*, not the scratches themselves, that causes the glass to fracture. I much prefer a new clause or sentence to a participial phrase or a *which* clause.

On the other hand, watch out for a construction like

> Although good practice requires periodic checks, this is not usually done.

Here, you may well ask, What is not usually done? Good practice? The pronoun *this* has no antecedent and cannot really be taken to refer to the entire clause that precedes it. This sentence should be rewritten

> Although good practice requires periodic checks, periodic checks are not usually carried out

or, perhaps,

> Although good practice requires periodic checks, these are not usually carried out.

This author was probably afraid to use the term *periodic checks* twice in succession; see **false elegance**.

this particular. This. If you really want to stress this particular one, you will have to avoid the badly overused phrase *this particular* and instead use *this one, in particular*.

those kind of things. That kind of thing. In this and similar expressions, the writer is evidently getting confused by the plural object of the preposition and attaching the plural *those* to the singular *kind.* He might mean *those kinds of things*, but usually *that kind of thing* suffices. See also **collective noun; singular or plural**.

though. See **although**.

thus. A synonym for *so* and, colloquially, for *therefore.* In technical writing, however, it is better to use *therefore* when that is what you mean. Incidentally, *thus* is an adverb (never an adjective), and it is incorrect to add **-ly** to it. See also **first, second**.

tick. A short line or mark, as in *ticking off* the items in a list. A short line that indicates the scale in a graph may be called a *tick* or a *hatch*, but it should not be called a *tick mark* and especially not a *tic mark* (a *tic* is a nervous twitch). See also **crosshatched**.

till. A perfectly good synonym for *until.* No need to write it with an apologetic apostrophe, *'til*, either. But it may be a little too colloquial for a formal paper.

time frame, time horizon, time window. Clumsy synonyms for **duration** or *period.* For example,

> We had only a 10 ns time window between the pulse
> and the reflected wave

could be rewritten

> We had only a 10 ns period between the pulse and
> the reflected wave.

This form is shorter and sounds less like **jargon**. *Time period* and *period of time*, incidentally, are redundant and should be just *period.* As for *time delay*, what other kind of delay could there be? Use *delay*.

title. The title is a very important part of your paper. More people see the title than even the **abstract**. The title should therefore be explicit and informative. It should not, on the other hand, be too long or wordy. The title needs only to tell what the paper is about, so that those more interested can decide whether to read the paper or just scan the abstract.

> Single-transverse-mode double-pulsed Q-switched ruby laser for high depth-of-field holography: theory and experiment

is not a title; it is an abstract. It could easily be shortened to

> High coherence ruby laser for double-pulse holography

without discouraging anyone from looking up the complete article. (The longest title in my notes, incidentally, is *Conditional symbolic modified single-digit arithmetic using optical content-addressable memory logic elements: Conditional symbolic modified signed-digit arithmetic operators*. I cannot edit this title, because I do not know what it means.)

Many editors ask you to begin your title with a key word, not with some dead phrase like *Analysis of* or even *Theory of*:

> Theory of scattering by polished glass surfaces.

Some journals will edit titles that begin with such phrases. If you want to tell explicitly that your paper is theoretical, you can append the word *theory*,

> Scattering by polished glass surfaces: theory,

or write it into the title,

> Perturbation theory of defects in waveguides.

Sometimes, though, a title has better style if it does not begin with a key word; for example, I cannot easily improve on

> Dynamics of laser-induced breakdown in gases,

perhaps because *dynamics* is a more descriptive word than *experiment* or *study*. See also **adjective; capitalization; dead words**.

titles. When referring to colleagues by their full names, as in **acknowledgements**, avoid using their titles, *Dr., Prof.,* and so on. If you insist, though, please use titles for everyone in the list; that is, do not write

> I am indebted to Dr. M. Richard Mendelson,
> Dr. Lowell Yemin, and Art Singer for their help.

Whether or not he has a Ph.D., Art Singer is entitled to as much respect as the others, and he has a title too. Write

> I am indebted to Dr. M. Richard Mendelson,
> Dr. Lowell Yemin, and Mr. Arthur J. Singer
> for their help.

In text, you may occasionally have to refer to colleagues by their last names. Except in certain publications like *The New York Times*, it has become customary to omit titles entirely:

> Shoemaker thought that the idea was not original,

not

> Mr. Shoemaker thought that the idea was not
> original.

In any case, do not follow the archaic custom of omitting titles for men but using them for women. That is, do not refer to a man as *Johnson* and to a woman as *Miss Johnson*, *Ms. Johnson*, or *Dr. Johnson*. The correct title for most women who do not have advanced degrees is *Ms*.

to be published. If you cite a reference as *to be published*, please be sure that you are really going to publish it. That means that the manuscript must be nearly ready or actually submitted. State the title and the name of the journal to which you have submitted the paper; otherwise the citation is useless. If you are not the author, be certain you have the authors' permission to cite their paper or to quote their data. See also **acknowledgements; private communication; references**.

to within. Using two prepositions in a row like this is usually unnecessary.

> The values agree to within 5 percent

is just as clear written

> The values agree within 5 percent.

tone. Flesch says to use a lot of personal words like *you* and *I*, and a lot of apostrophes, as in *don't* or *can't*. You can do that if you want to, but in technical writing you often won't establish the right tone. You want your style to be serious, though not solemn. I try to write as colloquially as possible while still keeping a scholarly tone when appropriate. In a technical paper, I would never use *don't* and rarely use *you*, but I refuse to give up *I*. For a **conference proceedings** paper, however, where it is clear that the paper is (more or less) a transcript of my spoken words, I will use a somewhat more colloquial style.

Once I was accused of oscillating between descriptive and pedagogical. Although I had not thought of it that way, that was exactly the style I was trying to achieve: to be descriptive when I was describing my experiment and to be pedagogical when I was trying to explain something. See also **first person**.

topic sentence. See **paragraph**.

tortuous, torturous. A *tortuous* path, not a *torturous* path. *Tortuous* means twisted or crooked and is related to *torque*. I have never seen *torturous* used correctly, but my dictionary assures me that it means full of torture. It is easy to understand why the author of the sentence,

> In a torturous bit of logic, the judge declared that AIDS is not a venereal disease,

may have felt tortured, but the judge's logic was *tortuous* (or *twisted*) not *torturous*.

toward. In writing, at least, *toward* is more common than *towards* and is the preferred U.S. usage.

trade names. Avoid trade names, except when you have a good reason to use them. Instead, use generic names whenever possible. Using a trade name, like Lucite or Plexiglass, when you mean *acrylic plastic*, is close to **jargon** and adds nothing to your paper. If, however, one manufacturer's product behaves significantly differently from other, similar products, your audience may need to know which one you used, for example, to duplicate your experiment. Even so, I suggest that you use the trade name once only, perhaps in parentheses or in a reference or footnote, and use the generic name thereafter.

Sometimes, however, a generic name is complicated or relatively little known. This is so with products like Pyrex or Teflon, whose names should not be regarded as generic names. Using these trade names is practically unavoidable; they should be capitalized as a courtesy to the manufacturers. You are, however, under no obligation to use the registered trademark symbol when you are referring to someone else's products.

transitional devices. To get from one sentence, one paragraph, or one thought to another, you often need some kind of transitional device to announce a shift or to avoid sounding repetitious. These include *also, yet, alternatively, thus, however, therefore, incidentally, otherwise; in addition, on the other hand, for example, as well as*; and, if they are used correctly, *and* and *but*. See, however, **and (comma), but (comma)**.

Most of these words and phrases are nonrestrictive: the sentence can be understood without them. Do you enclose them with commas? One school of thought says not to, because they are so short, but some journals almost always set these phrases off with commas. I think, however, that you can afford to be somewhat inconsistent and use commas depending on the use and position in the sentence. For example, I might write

> I therefore conclude...

but

> I conclude, therefore, ...

Read the sentence out loud. With *therefore* in the first position, you would not pause for it, but in the second position you probably would. But, then, I use somewhat more commas than many writers.

trivial. When you have worked hard at something and understand it, it becomes trivial to you. Never assume it is trivial to someone else. Your job as a writer is to explain it, not to gloss over it or tell me it is obvious. See **Baumeister's law**.

true fact. All facts are true; if they were not, they would not be facts. Thus, you do not check someone's *facts*; you check his *statements*. (Incidentally, all confessions are also true; otherwise, they are lies.) See also **factoid; redundant expressions**.

try to. Use *try and* only to indicate the passage of time between the trying and the doing. That is, you don't *try and* do something; you *try to* do it. You can, however, try *and* fail.

two cultures. C. P. Snow was right when he wrote that there are two cultures: the scientists and engineers, and the humanists, by which I suppose he meant everyone else. There are, however, those who belong to both cultures, and virtually all of them are scientists and engineers. (Many scientists can play a violin; very few musicians can solve a differential equation.) It is therefore the scientists' job, not the humanists', to bridge that gap, in part by learning to communicate clearly, to fellow scientists and nonscientists alike.

type, -type. I suggest a moratorium on using *type* as a noun, either on its own or as a suffix, except when it has to do with printing. Two particularly bad examples I ran across are

> confirmatory-type experiments

and

> pencil of the china-marking type;

these should have been simply

> confirmatory experiments

and

> china-marking pencil.

Often *-type* makes otherwise good writing look like **jargon**. For example,

> projection-type three-dimensional holography

is neither as good nor as clear as

> three-dimensional projection holography.

The more natural **word order** also appears when *-type* is omitted.

These masses are orbiting several solar-type stars

is not just clumsy; it is wrong. The author meant to say

These masses are orbiting several sun-like stars.

On the other hand,

This wild-type bacterium produces an antifungal agent

is unclear. Is *-type* gratuitous, or does the author really mean a bacterium that is similar to a wild strain? If the latter, the sentence should have been

This modified wild bacterium produces an antifungal agent

or something similar. See also **keyboarding**.

typical. Use *typical* only when you really mean it. Often

Figure 3 shows a typical index profile

translates into

Figure 3 shows our best result.

See also **preliminary experiment**.

U

un-. See **de-, un-; non-; prefix, suffix**.

under way. Two words.

unique. Not a synonym for *very unusual.* Something is *unique* only if there is nothing else like it; it can't be *very unique. Unique* is an **absolute word** and should not be intensified.

units. See **SI units; symbols**.

unity. You don't write *equal to two* or use the term *duality*, and you never say *unity*. Then why write *unity* when you mean 1? Or **zero** when you mean 0? These practices were begun, probably, to avoid confusion with lowercase *ell* and uppercase *oh.* Writing *zero* is perhaps defensible, but *unity* is always stilted. Usually context will resolve any ambiguity; if it doesn't, rewrite the sentence.

Another ploy to avoid writing 1 is to write 1.0. Strictly speaking, this is wrong; 1 means precisely 1, *the digit 1*, whereas by convention 1.0 means *a number between 0.95 and 1.05*. Write 1 when you mean 1.

unloosen. This ought to mean *tighten*, but it means *loosen*. It's a poor synonym, so stick with *loosen.* But do not confuse with *unleash*, which means *release*, but has a stronger connotation.

up to. Reserve *up to* for hypothetical situations. For example, write

> There may have been up to 1000 ions in the trap,

or

> There were nearly 1000 ions in the trap,

but not

> There were up to 1000 ions in the trap.

See also **as many as**.

utilize. There is not one iota of difference between *utilize* and *use* (the verb). Why then use the clumsy *utilize*? Use *use*.

The noun *utilization* is worse than *utilize* and should always be replaced by *use*. *Utility* is exactly synonymous with *usefulness*.

V

vague words, vogue words. Sometimes difficult to distinguish. A *vague word* is a word that does not have a precise meaning or is not precise enough for a given application. Typical vague words are *nice, fine, really, good, bad*, and, in many contexts, *important, kind, type, character, relative, appreciable*, or *device*. If you use an adjective, for example, but always have to intensify it with a word like *actually, very, totally, extremely*, or *exceptionally*, drop the intensifier or look for a more precise adjective that means what you want to say.

A word like *apparatus* or *device* can also be a vague word. For example, do not write

> We used a device that dispersed the light into its constituent colors.

In this context, *device* is too vague; tell us what kind of device:

> We used a prism spectrometer to disperse the light into its constituent colors.

Words like *imperative* and *essential* (and their adverbs) are *vogue words*, that is, words that are in style and have lost much of their strength. For example, *essential* does not mean just *important*; it means *of the utmost importance*. If you have to intensify it with *extremely*, you are using a vogue word that has lost much of its force. See whether *important* will suffice, or try to find another word, such as *urgent* or *indispensable*, that more nearly fits your need.

Other vogue words are technical words clumsily applied to other fields. These include *parameter, bottom line*, **interface**, *mode*, and **space**; phrases like *immediate* **feedback** and *close the loop*; and, in a sense, *ballpark figure*, and *touch base with you*. The suffixes **-wise** and **-type** are also in vogue. Avoid these words except in fields where they are precisely defined. For example, a *mode* is a field distribution; don't say (much less write)

> He was in a defeatist mode that day

unless you are doing it for laughs.

Adjectives like *appreciable, relative*, and **reasonable** are also vogue words and may be vague. *Relatively large* in microscopy is not even significant in astronomy. Therefore, don't write

> The surface feature was relatively large,

but, rather, something like

> The surface feature was several kilometers across

or

> The surface feature was easily resolved by the telescope.

See also **relative to**.

Adverbs can similarly be vogue words. Some of these are the adverbs used to intensify adjectives that have become weak because they are in vogue: *absolutely essential, totally negligible*, or *precisely similar*, for example.

Vogue words like *field* (as in *the field of technical writing*), *area*, and *character* can contribute to **circumlocutions** or **wordiness**. Just one example:

> Noise of the stationary or random character was also studied as a function of aircraft tested

should have been shortened to

> Stationary, or random, noise was also studied for each aircraft.

Process is another vogue word. In

> We are in the process of measuring the viscosity,

delete *in the process of*. See also **dead words**. Likewise, *processing* is a vague but voguish word. Even though it might have been clear in context,

> Algorithms for processing are more important than the attention paid to them

should have been more specific:

> Algorithms for optimizing clock readings deserve more attention than they have received.

Basis can similarly be used as a vogue word:

> Different values are tried on an iterative basis until a uniform solution is found

should be rewritten

> Different values are tried iteratively until a uniform solution is found.

Finally, **unique** has become a vogue word and is losing its unique meaning; see **absolute words**. See also **seem; weasel words; would**.

various. Overused, especially with the definite article, *the*. Do not write

> The various voltages were fed to the voltmeter by a multiplexer.

Leave out either *the* or *various*, depending on your meaning:

> The voltages were fed to the voltmeter by a multiplexer

or

> Various voltages were fed to the voltmeter by a multiplexer.

In the first revision, the voltages (which we have just discussed) were fed to the voltmeter, whereas, in the second revision, various voltages in general were fed to the voltmeter.

Sometimes, *various* is used when *different* is called for. For example,

> The various averaging techniques give the same functional form

really means

> These different averaging techniques give the same functional form.

vary. On TV, when they say

> Prices may vary,

they really mean

> Prices may differ in different places

or

Prices may change without notice.

Vary implies irregular or random change and should be used only in a sentence like

The temperature varied during the experiment.

This has a very different connotation than

The temperature changed during the experiment.

In the first example, the temperature varied erratically, whereas, in the second example, it changed from one constant value to another.

vee. See **tee, vee**.

verbifying. Making a verb out of a noun, either by adding **-ize** or by just using the noun in place of a verb. *Formularize* is a verbification, as is *impact on* (for *influence*). In the sentence,

Ultrasound breaks up the cataract, and the lens is suctioned out,

there was no need to coin the verb *to suction*. The sentence should have been

Ultrasound breaks up the cataract, and the lens is sucked out.

Avoid verbifying, and use only recognized verbs.

versus. Using *versus* or its abbreviation, *vs.*, can lead to a clumsy expression like

the microwave-attenuation-versus-frequency plot.

It is easier to read

a plot of microwave attenuation versus frequency,

but

a plot of microwave attenuation as a function of frequency

is perhaps best. *Versus*, which is Latin for *against*, sounds like

jargon and does not really convey the dependence of one variable on the other. See also **foreign words and phrases**.

via. Most properly pronounced with the long *i*, this preposition means *by way of* or *through*. It should be reserved almost exclusively for planes, trains, and ships, and is a poor synonym for *by*, *by means of*, or *from*. For example,

> They performed the experiments via a video frame-digitizing device

should be rewritten as

> They performed the experiments by means of a video frame-digitizing device

or

> They performed the experiments with a video frame-digitizing device.

On the other hand,

> The information is disseminated via satellite

seems OK because the information is transmitted from the ground to other locations *by way of* a satellite. If you can't substitute *by way of* for *via*, find some other preposition. See also **vague words, vogue words**.

viewgraphs. See **slides, viewgraphs**.

visible laser. All lasers are visible, but some emit invisible radiation. Nevertheless, I would defend this phrase, which is really shorthand for the clumsy *visible light laser*. (In the same way, incidentally, all mushrooms are edible, though some will kill you if you eat them.)

visible light. Nowadays, we define light to include a far-greater portion of the electromagnetic spectrum than merely the visible portion. *Visible light* is not redundant, since infrared light or ultraviolet light is invisible light.

voluntarism. Not voluntEERism.

W

wait time. Waiting time.

wastebasket. Don't use it hastily. Save your false starts and discarded paragraphs. You may want to return to them, and it's hard when they are crumpled up and covered with coffee or glue. Throw something out only when you are sure you are satisfied with its revision.

Incidentally, this points out one problem with word processors: Sometimes you are apt to erase something and then later be unable to remember it accurately. It may be worthwhile to save revisions in temporary files or even print short segments before discarding them. One device is to have two files (or two disks) and save successive revisions alternately on each file. This way, you can often recover something, rather than rack your brain trying to recall something you had already consigned to the wastebasket in your head.

watt of power. The watt is the unit of power, so you can't have a watt of anything else. The phrase is redundant and should be simply *watt*. You can, however, say *a power of 10 watts* and maybe even *10 watts of electricity* without redundancy.

One of the more common errors in use of units involves watts and energy. The sentence

Waves deliver 0.335 W/cm^2 of energy to the coastline

is incorrect, not just because three significant figures are probably too many, but also because the unit of energy is not the watt per square centimeter. This kind of error is most often made in popular articles, but sometimes it appears in technical papers as well.

weight. See **mass, weight**.

we. Don't say *we* if you are the only author and are referring to yourself; it always seems pompous. An exception is when you are writing a report, as from one agency to another, that does not have a byline like a technical article. When I read a paper in which a man constantly refers to himself as *we*, I always want to ask, How do you refer to your wife? See also **first person**.

weasel words. Words that some people use to be evasive or to avoid having to make definite statements. Typical weasel words are **would, seem**, and, sometimes, **hopefully**. These allow the writer to equivocate or, at least, give the appearance of equivocating.

Ambiguous expressions and **euphemisms** can also be used as weasel words. These are more common in advertising than in science, but a phrase like **preliminary experiment**, for example, may be weasel words. Likewise, what does it mean to say

> To study this concept theoretically was somewhat intractable?

Either the problem was intractable or it was not; you do not need the weasel word *somewhat*. See also **vague words, vogue words**.

Avoid weasel words; make definite statements when they are called for, and draw definite conclusions.

well. Except when it means *not sick* or *a hole in the ground*, *well* is an adverb. I argue that compound words with *well* need not be hyphenated: since *well* is always an adverb, there can be no redundancy. Dictionaries, on the other hand, hyphenate words like *well-known, well-defined*, and *well-founded*. Other authorities would hyphenate these words when they precede the noun they modify, but not when they are used as predicate adjectives:

> The fact was well known at the time,

but

> It was a well-known fact.

When writing for a publication that will edit your paper, therefore, hyphenate *well* words according to the publication's style. In camera-ready material, do it however you prefer, but be consistent.

See also **adverb; good; -ly**.

well known. Usually part of a wasted phrase, as in

> It is well known that stresses exist in films and may be large enough to cause thick films to delaminate.

The author usually means *widely known*, rather than *well known*. (Something can be well known to a few, but not widely known.) If something is widely known, you don't have to tell us so; just tells us what you want to say. If we know it already, we know it already. The sentence would be best rewritten

> The stresses that exist in films may be large enough to cause thick films to delaminate.

See also **Baumeister's law; dead words; it**.

what are, what is. The sentence

> The community struggles with the conflict between the unbendable, from its point of view, demands of the law and the real claims of life in the present

could profit by addition of the phrase *what are*:

> The community struggles with the conflict between what are, from its point of view, the unbendable demands of the law and the real claims of life in the present.

What is can also be a less stilted synonym for **that which**.

whence. See **thence, whence**.

where's it at? Where is it? You would not say, *At where is it?* since *where* means *at what place*, so why say, *Where's it at?*

whereas. Although a bit formal, *whereas* is often a better word than *while*, which is sometimes used imprecisely. For example,

> While we were unable to measure position, we measured velocity

could mean

> During the time that we were unable to measure position [later we succeeded in measuring position], we measured velocity,

or it could mean

> Whereas we were unable to measure position, we measured velocity.

Most likely the intended meaning is *whereas*. Thus, although it is a somewhat formal word, *whereas* should be used in order to avoid an ambiguous or incorrect interpretation.

whether. Often better than *if* in a sentence like

> We do not yet know if the theory works,

where you are expressing doubt. This sentence would be slightly clearer

> We do not yet know whether the theory works.

You do not usually need to add *or not* to *whether*, but if *whether or not* works in your sentence, use *whether* in place of *if*. Use *whether or not* only when you especially want to point out the element of doubt. See also **if; may or may not**.

which. See **misplaced modifier; this; that, which**.

while. See **whereas**.

who. Use the pronoun *who*, not *that* or *which*, when referring to people.

> There are 3100 employees, of which 600 are engineers

should have been

> There are 3100 employees, of whom 600 are engineers.

whom. There was a proposal in a national magazine a few years ago to eliminate *whom* except after a preposition: "I favor whom's doom, except after a preposition." The argument is very compelling. Hardly anyone uses *whom* "correctly" any more, and sometimes it sounds pompous, as in

> I wondered whom he was asking.

In addition, you will sometimes hear *whom* or *whomever* used where *who* or *whoever* is correct:

> He looked for whomever he thought could help him.

Here *whoever* is correct, because it is the subject of the noun clause *whoever could help him.* That clause, not the pronoun *whoever*, is the object of the preposition *for*. Unfortunately, either the nonrestrictive phrase *he thought* or the preposition itself seems to demand the objective case *whomever.*

We have few case endings in English anyway; why not throw away *whom* except as the object of a preposition?

width. Obsolete synonym for **duration**.

window of opportunity. Spare me. Delete *window of.*

-wise. In the direction of or in the manner of. *-wise* is OK in standard phrases like *clockwise* or *piecewise continuous.* But don't tack it onto every conceivable noun and use it to mean *from the viewpoint of*, as in

> Equipmentwise, we are well supported.

Just say

> We have good equipment.

See also **vague words, vogue words**.

without further ado. Cut this **cliche** without further ado.

woman scientist. Some people get upset when they hear this phrase. It is, however, difficult to write an article about, say, the increasing importance of women in science without using such a phrase to distinguish between scientists who are women and those who are not. (Some would argue that *female scientist* is correct and avoids the ambiguity of *woman doctor*, but the accepted phrase is *woman scientist.*)

What is the plural of *woman scientist*? In English, adjectives do not have to agree with their nouns in number (singular or plural), simply because we do not put endings on adjectives. Neither do nouns when they are used as adjectives. The plural of *computer scientist* is *computer scientists.* We say *footballs*, not *feetballs.* Why then do we say *women scientists* instead of *woman scientists*? See **Ciardi's law**.

word order. Word order is usually obvious to the native speaker. We say *a big, red balloon*; *a red, big balloon* is obviously wrong.

In scientific writing, a group of words sometimes forms a unit in the mind of the writer. Hence you sometimes see a construction like *an argon, single-mode laser*. The author was apparently thinking *single-mode laser* and then tried to modify it with the noun, *argon*. The word order, however, is unusual. The more natural word order is *a single-mode argon laser*.

What if you have a pure fluid that you want to use as a standard or reference material? Is it a *reference pure fluid*? I think not; the more natural word order is *pure reference fluid*. Using unnatural word order makes your writing clumsy.

Sometimes a phrase becomes standard even though its word order is incorrect. There is a difference between *ice water* and *water ice*. *Water ice* means frozen water, not frozen methane, for example. Obvious as this may be, every field of science has expressions like *gradient index optics*, which should logically be *index gradient optics*, since it is the optics of materials that have index gradients; *slush hydrogen*, which should be *hydrogen slush*, by analogy with *water ice*; and *corner cube*, which should be *cube corner*, since it is the corner of a cube. In my writing, I try to use the right word order, even if it is not standard, though I usually mention the incorrect word order and include it in my list of key words, if appropriate.

The position of an **adverb** in a sentence is often not critical. See, however, **only; split infinitive**.

For a discussion of *reverse word order* and the order of the subject, verb, and object in a sentence, see **passive voice**. See also **dangling modifier; misplaced modifier**.

wordiness. You can be wordy by using long sentences with lots of subordinate clauses, by piling modifiers on top of one another, by using **dead words**, and probably by using many other methods that have not been discovered yet. Much of the thrust of this book is to avoid wordiness. Here I will give some examples of wordy sentences, some of them not so long, and their concise revisions—what an editor called *a list of edited monstrosities*. They are presented in no special order.

Wordy:

> The intensity is diminished from that which would be observed over a closed path.

Concise:

The intensity is less than what would be seen over a closed path.

Wordy:

This program has the capability to select the number of operations.

Concise:

This program can select the number of operations.

Wordy:

It may be that some of this behavior is learned.

Concise:

Some of this behavior may be learned.

Wordy:

The steady-state thermal-conductivity guarded-hot-plate and heat-flow meter apparatus require stabilities in the 10 mK range.

Concise:

The guarded hot plate and heat flow meter for measuring thermal conductivity in the steady state require stabilities in the 10 mK range.

Wordy:

Some approximation of common sense is absolutely critical.

Concise:

Common sense is critical.

Wordy:

It is expected that Case 3 will yield similar results.

Concise:

Case 3 will probably yield similar results.

Wordy:

> The technique of laser cooling utilizes the resonant scattering of light by atomic particles.

Concise:

> Laser cooling uses resonant scattering of light by atomic particles.

Wordy:

> We choose to use convolution to take advantage of its simpler implementation and faster computational time with convolutions which require small kernels.

Concise:

> We choose convolution because it is simpler and faster with small kernels.

Wordy:

> Among hip-fracture patients, there is a mortality that may be as high as 12 percent.

Concise:

> Among hip-fracture patients, mortality may be as high as 12 percent.

Wordy:

> The vane shown in Figure 6, when paired with a vane like the bottom one, but without the sensor and the notch, which will be designated the stiff vane, produced the data shown by the circles.

Concise:

> The vane shown in Figure 6 was paired with a vane like the bottom vane, but without the sensor and the notch. It is called the stiff vane and produced the data shown by the circles.

Wordy:

> Once the parameters are established, it would seem reasonable to achieve repeatability with this process.

Concise:

> Once the parameters are established, the process will probably be repeatable.

Wordy:

> A general property of the critical current of these bulk samples is that their critical current is low.

Concise:

> In general, the critical current of these bulk samples is low.

I will conclude this section with a sentence from a scientist's autobiography. It is so complex and wordy that I cannot understand it and therefore cannot generate a concise version. Doubtless it was clear to its author.

> Like my colony of glow-worms (larvae of the common firefly) that secreted a digestive enzyme to extract some nourishment from a piece of boiled potato, earthworm, or a cricket that, not knowing what to feed them, I provided for food, I can envision the various "ists" trying to extract some juice from our records and then attempt to correlate our various moods, tenses, and fits of cussedness.

would. *I would hope, I would suggest, I would think.* No. You *do* hope, you *do* suggest, and (I hope) you *do* think. Why water your verbs down with the extraneous and, as they say (not would say) today, wimpish auxiliary verb *would*? Likewise, *would seem* and *seems to suggest* are overly qualified. In all these, drop *would* and *seems.*

> I would hope that the treatment would be appealing to engineers

is far better as

> I hope that the treatment is appealing to engineers.

Would is also used incorrectly in the conditional,

> If he would have gone, I would have gone too.

This should be

If he had gone, I would have gone too.

On a more technical note,

One consequence would be that modern geographic populations would have deep roots

probably should be

One consequence would be that modern geographical populations have deep roots

or possibly, depending on context,

One consequence is that modern geographic populations have deep roots.

See also **seem; weasel words**.

writer. No good writer ever calls himself *the writer*. See **first person**.

X Y Z

x number of things. See **few number**.

you. Sometimes it is appropriate to use *you* in a technical paper, for example, to replace **one**. For example,

> If a bottleneck occurred in this manner, how would one know?

reads better as

> If a bottleneck occurred in this manner, how would you (or we) know?

On the other hand, it would be hard to improve on

> Ask yourself whether you would buy new spectacles with a scratch across the lens.

You is also used implicitly in a sentence like

> Take [or consider] Fresnel zone plates as an example.

See also **first person**.

you knew what I meant, dincha? What University of Colorado students say when you correct their writing in an engineering course. Many other people feel the same way: if you can understand them, that is good enough. Well, maybe it is, if you do not have to make a special effort to understand them. But often I understood only because I had already done the problem.

Do not be satisfied if you write so that someone can decipher it. Tell your story so that anyone with the right background can understand and even enjoy it.

you know. Bremner spells this *yunno* and calls it a *knee-jerk word* because you, like, you know, sort of say it, um, I mean, reflexively, without thinking about it. Bremner's impatience may not be justified, but some people say *you know* so much that it is more like a verbal tic than a knee jerk. An occasional *um-m-m-m* may be OK to buy a little time to answer a question, but, especially in a formal talk or lecture, try to avoid these knee-jerk words. If you think your audience might not understand, just ask.

zero. There is no need to spell *zero*; you can almost always use 0 without being misunderstood. An exception is in a phrase like *absolute zero, zero temperature,* or *zero energy,* where *zero* almost has to be spelled fully. But when the number 0 stands alone and does not function as an adjective, write

The temperature approaches 0,

rather than

The temperature approaches zero.

See also **naked decimal point; numbers in sentence; unity**.

Bibliography

The American Heritage Dictionary, second college edition, Houghton-Mifflin, New York, 1976.

American National Standard for the Preparation of Scientific Papers for Written or Oral Presentation, ANSI Standard Z39.16-1979, American National Standards Institute, New York, 1979.

Ron S. Blicq, *Writing Reports to Get Results: Guidelines for the Computer Age*, IEEE Press, New York, 1987.

John B. Bremner, *Words on Words*, Columbia University Press, New York, 1980.

Charles T. Brusaw, Gerald J. Alred, and Walter E. Oliu, *Handbook of Technical Writing*, second edition, St. Martin's Press, New York, 1982.

The Chicago Manual of Style, thirteenth edition, University of Chicago Press, 1982.

Effective Lecture Slides, Eastman Kodak Company Publication S-22, latest edition.

Rudolf Flesch, *The ABC of Style*, Perennial Library, New York, 1980.

Rudolf Flesch, *The Art of Readable Writing*, Harper, New York, 1949.

Karen E. Gordon, *The Transitive Vampire*, Times Books, New York, 1984.

Karen E. Gordon, *The Well-Tempered Sentence*, Tichnor and Fields, New York, 1983.

David Hathwell and A. W. Kenneth Metzner, *Style Manual*, American Institute of Physics Publication R-283, New York, 1978.

Porter G. Perrin, *Reference Handbook of Grammar and Usage*, William Morrow, New York, 1972.

Roget's International Thesaurus, fourth edition, Thomas Crowell, New York, 1977.

William H. Strunk, Jr., and E. B. White, *The Elements of Style*, third edition, Macmillan, New York, 1979.

Henrietta J. Tichy, *Effective Writing for Engineers-Managers-Scientists*, Wiley, New York, 1966.

Christine Timmons and Frank Gibney, editors, *The Britannica Book of English Usage*, Doubleday Britannica Books, Garden City, New York, 1980.

Edward R. Tufte, *The Visual Display of Quantitative Information*, Graphics Press, Cheshire, Connecticut, 1983.

The United States Government Printing Office Style Manual, Superintendent of Documents, United States Government Printing Office, Washington, DC, 1984.

Webster's Collegiate Thesaurus, G. and C. Merriam, Springfield, Massachusetts, 1976.

Webster's New Collegiate Dictionary, G. and C. Merriam, Springfield, Massachusetts, 1977.